YOU & ME

THE SKILLS OF COMMUNICATING AND RELATING TO OTHERS

OTHER BOOKS BY GERARD EGAN

ENCOUNTER
Group Processes for Interpersonal Growth

FACE TO FACE
The Small-Group Experience
and Interpersonal Growth

THE SKILLED HELPER
A Model for Systematic Helping
and Interpersonal Relating

EXERCISES IN HELPING SKILLS
A Training Manual to Accompany
The Skilled Helper

INTERPERSONAL LIVING
A Skills/Contract Approach to
Human-Relations Training in Groups

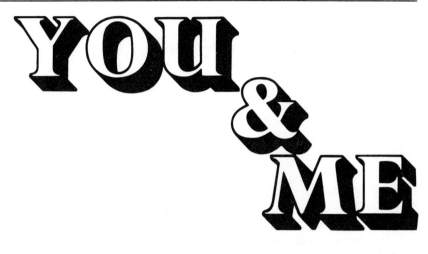

YOU & ME

THE SKILLS OF COMMUNICATING AND RELATING TO OTHERS

Gerard Egan

Loyola University of Chicago

BROOKS/COLE PUBLISHING COMPANY
MONTEREY, CALIFORNIA

A Division of Wadsworth Publishing Company, Inc.

I**T̂P**™ The trademark ITP is used under license.

Printed in the United States of America

21 22 23 24

Library of Congress Cataloging in Publication Data

Egan, Gerard.
    You and me.

    Bibliography: p. 335
    Includes index.
    1. Interpersonal relations. 2. Interpersonal communication. I. Title.
HM132.E34   158'.2   77-6475
ISBN 0-8185-0238-X

Production Editor: *John Bergez*
Cover and Interior Design: *Linda Marcetti*
Chapter-Opening Photos: Chapters 1, 10, and 12, *Gerard Egan;* Chapters 2, 7, and 9, *Helen Nestor;* Chapters 3, 6, and 13, *Central YMCA Community College, Chicago;* Chapter 4, *Jim Pinckney;* Chapter 5, *David Glaubinger and Jeroboam, Inc.;* Chapter 8, *Karen Preuss and Jeroboam, Inc.;* Chapters 11 and 14, *Russell Abraham and Jeroboam, Inc.*

Preface

This book is intended for people who wish to improve their interpersonal self-awareness, skills, and assertiveness. It is designed especially for those who wish to pursue these goals through participation in a small-group experience.

It would be ideal if a combination of family living and primary school experiences equipped us with the communication skills we need to involve ourselves intimately and creatively with others, but, unfortunately, that doesn't seem to be the case. If we want these skills, we have to work at acquiring them.

This book presents a step-by-step system for:

- deepening our awareness of ourselves as social or interpersonal beings,
- developing the skills we need to make our interpersonal involvements richer, and
- helping us to become more outgoing or active interpersonally.

Why such a *systematic* approach? Studies have shown that we don't necessarily develop good communication skills merely by being with and talking to others. Learning these skills in a systematic, step-by-step way is simply more efficient.

This book is written for the lay reader. It doesn't use technical language, nor does it presuppose any previous study of psychology. (For a more technical treatment, see my *Interpersonal Living*, Brooks/Cole, 1976.)

I would like to thank the following people for reading the manuscript and making suggestions for its improvement: Dorothy Bushnell, Merced College (Merced, California); Dan Fallon, Central YMCA Community College (Chicago); George Pilkey, Fulton

Montgomery Community College (Johnstown, New York); Allen Segrist, Purdue University; and Jerry Wesson, El Centro College (Dallas). As always, the students in the training groups remain the richest source of my learning.

Gerard Egan

Contents

Contents

Checklists and Exercises

Responding with Understanding

Trust

Identifying Strengths

Deeper Understanding

Confrontation

Encouraging Immediacy in the Group

Promoting Effective Group Communication

Challenging Individual and Group Flight

Encouraging Openness in the Group

Changing Your Behavior

YOU & ME

THE SKILLS OF COMMUNICATING AND RELATING TO OTHERS

Human-Relations Training: What This Book Is About

Chapters 1 and 2 introduce the process of learning, practicing, and using interpersonal-communication skills. Chapter 1 provides background for experiential approaches to learning—that is, for learning by doing. The chapter also discusses the values that this book promotes.

We spend a great deal of our lives talking with other people. We talk with our families, our friends (and sometimes our enemies), our teachers, our fellow students, the people we work with, and many others. Since we spend so much time and effort trying to communicate with one another, it's strange that so little of our time is spent learning *how* to talk and listen to one another effectively. It's especially strange because an inability to communicate interferes with all levels of human living. Psychological studies have shown, for instance, that even good friends are often not very good at talking with each other about deeper things. Many of us are often at a loss to know what to say once we move beyond the weather and the results of sports events. And many married couples end up either with a marriage counselor or in the divorce court because they discover that they "can't communicate" with each other.

This book is an invitation to ask yourself: "Just how well do I communicate with the people in my life? Do I want to improve my ways of relating to others? If so, how can I go about it?"

Human Relations: A Skills Approach

In relating to or communicating with others, we use what are called *interpersonal skills* or *communication skills*. The dictionary defines the word *skill* as "the ability to use one's knowledge effectively and readily in execution or performance" or "a learned power of doing a thing competently." An interpersonal skill, then, isn't just knowing what good communication is; rather, it's actually performing well in interpersonal situations. Interpersonal skills give you *power*—not power *over* people but power to involve yourself with people competently.

Learning a skill is usually a step-by-step process. Take, for instance, the skill of writing. First you learn how to write clear and thoughtful sentences. Once you do this well, you go on to learn how to put such sentences together into meaningful paragraphs. Finally, if you have good ideas, you learn how to write good articles or stories. Learning interpersonal skills is also a step-by-step process.

In the next chapter there is an outline of the precise skills covered in this book. For instance, one of these skills is the skill of listening to others. This skill includes:

- paying attention to others;
- "hearing" what others say with their bodies, their gestures, their faces;
- understanding accurately what others are thinking and feeling;
- communicating to others that you do understand or that you're trying to understand them.

Notice that full listening involves more than just hearing what others have to say. Full listening includes responding to others, letting them know that you have listened and that you understand. Perhaps you aren't used to considering listening (something you do automatically every day) as a skill. But effective listening is a "learned power of doing a thing competently." Becoming a good listener, an *active* listener, is more than knowing what goes into good listening. As with other skills, good listening is acquired by practice.

Using This Book

There are basically two ways of using this book to learn skills: you can use it alone, or you can use it in a group.

You could read the book, do some of the exercises, and then try to practice what you learn in your conversations with your friends, your family, your fellow workers, and the others you meet in your day-to-day life. In this case the book itself would be your instructor.

On the other hand, as you've probably already guessed, there is a better way of using this book, and that way is to learn these skills with a group of people under the direction of an instructor. Furthermore, your learning will probably be more effective if it's *systematic*. By "systematic" I mean learning one skill at a time by means of a step-by-step process. The process I have in mind includes five steps: (1) instruction, (2) practice, (3) feedback, (4) encouragement or support, and (5) the use of your new skill outside the classroom or learning group. Let's look at each of these separately.

1. *Instruction.* First you'll read a section of the book. Then your instructor will answer any questions you may have about what you've read. It's important to have a clear understanding of each skill before trying to practice it. Next, the instructor will give an example or a demonstration of each skill. Once you understand the skill and see it demonstrated, you might be asked to try the skill out yourself by means of a written exercise. Written exercises will help you get a better understanding of the skill, and they will also give you a chance to rehearse the skill in private. Written exercises are a link between instruction and practice.

2. *Practice.* Next your instructor will show you how to practice each skill with the other members of your group. As with any other skill (such as writing, playing ball, giving a speech), live practice is essential. The group is a safe place to practice, because it's a place where you can make mistakes and get help in correcting them. In the group you'll learn by doing, not by reading and memorizing.

3. *Feedback.* Feedback means being told by others what you're doing right, so that you can continue to do it, and what you're doing wrong, so that you can correct it. Your practice sessions will consist of conversations with your fellow group members. After practicing a skill in these conversations, it's important for you to find out how you're doing. Feedback means two things: it means that you tell others how well they're learning each skill, and it also means that you expect others to tell you how you're doing. For example, you might say to one of the other members of the group:

> "Mary, you understood what I was trying to say very well. I was feeling frustrated, and you picked that up and let me know that you could see my frustration. I felt that I had your full attention."

5

Or someone might say to you:

> "When you talk to me, you don't look directly at me very much. I don't expect you to stare me down, but, when you hardly look at me at all, you seem distant or uninterested in what I'm saying."

Feedback means that you tell another person in the group what his or her interpersonal behavior looks like to you and how it affects you. If feedback is given in a helpful, friendly way, and if it is honest and direct, it can be one of the most useful things that take place in the group.

4. *Encouragement or support.* Most of us need encouragement from others, especially if we're trying to do something difficult. As you'll probably discover, learning the skills outlined in this book is often hard work. Getting encouragement and support from both your instructor and your fellow group members is important. Too often people tell us when we do something wrong but don't say anything when we do something right. We also need encouragement when we try hard and fail. Maybe that's the time we need encouragement the most.

If you and your fellow group members are sincerely interested in one another's learning and are willing to provide encouragement for one another, then you have more than just a class or a skills-training group—you have a community, a *learning community*, even though it's only a temporary one. Effective communities provide both support and challenge for their members. Encouragement means more than patting one another on the back when things are going well. Giving caring, honest, and direct feedback—feedback that isn't meant to be harsh or punishing—is also a form of encouragement.

Finally, in a good community, cooperation rather than competition is encouraged. You're not in the group to compete with your fellow participants, to show them that you can learn interpersonal skills better than they can. If group members begin trying to show one another who is the best learner, the learning experience will be turned into a game, and an unpleasant game at that. Pursuing excellence in interpersonal skills rather than trying to be better than somebody else is a goal that encourages cooperation rather than competition.

5. *The use of interpersonal skills outside the training group.* This course or training experience is meant to be very practical. What you

6

learn here is intended to be taken out and used in your everyday life. Many people report an almost immediate payoff: comments like "I've begun to listen to people better, and I respond with much more understanding" are fairly typical, even near the beginning of the training experience. During group meetings you'll be given an opportunity to discuss your successes and failures in using your skills outside the training group.

The Group As a Laboratory

Sometimes human-relations training is called *laboratory learning*, and the group itself is called a *laboratory*. The reason for using these terms is to let people know that human-relations training is *different* from the kinds of learning that take place in most schools. In most classrooms the teacher talks, the students listen or take notes or ask questions, there are books to read and homework assignments to be turned in, and usually the students don't speak to one another very much (or at least they're not supposed to). Laboratory learning is very different. As you know, a laboratory is a place where scientists do experiments, trying things that haven't been tried before. Your group isn't exactly like a scientific laboratory, of course, but it is like a laboratory in a number of ways. For instance, in a laboratory people learn mainly by doing, not by listening to lectures. The learning process suggested in this book is a combination of laboratory learning and systematic training in skills. Let's see more concretely what this combination means.

Experimentation. Laboratory learning always includes the spirit of experimentation. Once you've learned basic interpersonal skills by being trained in them systematically, you'll use them in your conversations with the other members of your group. In a sense, you'll be experimenting with your interpersonal behavior, trying out new ways of acting with other people. For instance, right now you might find it very difficult to challenge or confront people. Or, if you do confront or challenge people, perhaps you do it poorly, with the result that they get angry with you and ignore you. Once you've learned how to challenge others in growthful ways (that is, once you've learned the skills of challenging), you can use these skills first in the group and then outside with your family and friends. Challenging others, then, would be a new way of acting for you. In the group you're given the opportunity to *experiment* with this skill until you feel

7

comfortable using it outside the group. You try it on, get the feel of it, and then use it. The group is a safe place to experiment with new skills and new ways of relating to other people. If you're shy, it's a safe place to learn how to speak up more. The feedback you receive in the group helps you to perfect your skills until you're ready to use them outside the group.

Anxiety. Since systematic skills training and laboratory learning are so different from ordinary classroom learning, they can make you feel somewhat uncomfortable or scared. While in an ordinary classroom it might be rather normal to be bored, in a human-relations-training group it's normal to be a bit scared. You can't learn skills without putting yourself on the spot, making mistakes, and trying again.

So, if you're somewhat fearful as you approach a group experience, you aren't different from most others. I was scared the first time I had to get up in front of a class to give a speech. I was also scared when I went to my first few human-relations-training groups. In fact, being a little scared isn't a bad thing. It keeps you on your toes. No one can learn good interpersonal skills without paying the price—working hard, failing and trying again, and feeling scared or on the spot at times. The kind of learning that can make you scared is also the kind of learning that can be exciting—a delightful change from more boring kinds of learning! In summary, then, if you do get anxious or scared, talk about it in the group. Talking about your feelings will help you lower your anxiety, and you'll also find out that a number of the other people in your group feel the same way. If your fear begins to get in the way of your learning, it becomes even more necessary to talk about it in the group. Sharing these kinds of concerns helps to make your group a learning community.

Emotional safety. Even though laboratory learning means learning how to take risks (for instance, even when you challenge or confront someone carefully, it takes guts to do it), it's very safe if the leadership (teachers, trainers) is good and if the members of the group are caring and encouraging. Systematic training provides another safety feature: you won't be asked to take a risk until you've been trained in the skills needed to take the risk. For instance, you'll learn that, to challenge another person sensibly, you must first take time to understand the person and to show him or her that you do understand. So the skill of understanding comes before the skill of challenging. Good leadership, plus systematic training in skills, plus encouragement and support from your trainer and your fellow group members, all add up to a high degree of psychological safety.

Are groups artificial? You'll be with your training group for a relatively short time, but during this time you'll most likely hear others talk rather deeply about themselves and find yourself doing the same. Isn't it awkward or artificial to get close so quickly to people who are relative strangers? Is this kind of laboratory learning phony or unreal in some way? In my experience, groups are only as artificial as their members make them. You're in the group not to establish deep friendships (although this may happen) but to establish relationships (we'll describe what this means in detail later on). You're in the group to learn skills and to practice them by developing relationships with your fellow group members—relationships that will last, usually, for only a short time. The group isn't a substitute for friendships and other relationships outside the group. Counselors have very real relationships with those who come to them for help, even though their clients don't become their friends and even though the relationships last for only a relatively short time. People who work together can have very good relationships with one another without becoming friends outside of work. The world is filled with intense, relatively short-term relationships that are nevertheless real. If you go into the group thinking that it will be artificial, it probably will be artificial for you, and you'll stand in the way of the other members' learning. But if you go into the group with an open mind and with a clear idea of what the goals of the group are—and these goals will be stated as clearly as possible in Chapter 2—then the group experience can be very real, and the relationships you establish can help you and your fellow group members to grow interpersonally.

There are two other possible sources of artificiality in this book: examples and exercises. First, many examples are used to clarify each skill presented. I've drawn these examples from my own experience with groups. Frankly, my experience with people may be different from yours. Therefore, since the examples are taken out of context, and since they're not necessarily representative of every kind of person, they might not always sound like the way people you know really talk. But keep in mind that the purpose of the examples is only to clarify the skill being presented. Your way of speaking may be different from somebody else's, but the interpersonal-communication skills used are the same.

Second, there are many exercises in this book. The exercises are a kind of rehearsal, a way of practicing skills by yourself before practicing them in the group. Since exercises aren't the same as real conversations, there can be something a bit unreal about them. Yet football teams, for example, are great only if they are great at *basics*. The players spend a great deal of time drilling on the practice field before they go out and play their games. In the same way, exercises, practice, drilling, and giving and getting feedback are all part of the price of achieving excellence in communicating with others.

On a more positive note, your group can be as real as you want to make it. The group meetings can be stimulating in themselves—a bright spot in the school week—and what you learn can be transferred to your relationships outside the group.

Values in Human Relationships

Doing the work that's necessary to improve your interpersonal skills is frustrating unless you see such improvement as *valuable*. So at this point you should ask yourself "Is relating to other people more honestly, more openly, more closely, more skillfully a value for *me*?" In order to answer this question, it's helpful to get a clear picture of just what a value is. Let's take an example. Someone asks you "Is helping your friends when they make reasonable requests for your help a value for you?" You might answer "Certainly." *But*, when you examine your behavior—what you actually *do*—you find that you seldom put yourself on the line and help your friends when they do ask you. Instead you find that you're too busy, that you have to study, that you're too tired, that you have a date; that is, you almost always have some excuse for not being able to help. Your friends finally get the message and stop asking you for help. Therefore,

- even though helping others seems to you to be a *good idea*,
- even though you might *feel good* when you see friends helping friends,
- even though your *attitude* toward helping others is positive, and
- even though you *wish* that you would help others more often,

still, helping others is not a *value* for you, because you simply don't find the time or energy or desire to actually do it. Good ideas, good feelings, positive attitudes, and good wishes don't add up to a value.

10

Something is a value for you only if you act on it, only if it becomes part of your *behavior*.

There are a number of reasons for bringing up the question of values at this point. Let's look at some of them briefly.

Values and motivation. First of all, if relating to others better isn't a value for you, you won't be motivated to do the work demanded by this training program. (On the other hand, the work will be fun if it's something you do want to do.) I used to assume that everyone wants to improve his or her way of relating to others, but my experience tells me that this isn't so. Someone in a course once complained to me that learning these skills was tiring work. I said to her that that was the price she had to pay if she wanted to live more deeply and closely with others. She came back the next week and said that, after thinking it over, she did *not* want to live more deeply and more closely with others. She was comfortable the way she was. She was honest with herself and with me, and her attitude explained why she was getting so tired doing the work required by the course.

Interpersonal values in action. A second reason for discussing values here is that your interpersonal values can be seen in the ways you relate to and act with other people. Part of this group experience involves taking a look at your *interpersonal style*. What does "interpersonal style" mean? If someone were to ask a friend of yours "What's he (she) like when he (she) is with people?" the answer your friend would give would be a description of your interpersonal style. Thus, "interpersonal style" means the usual ways in which you act toward people. For instance, if you talk deeply about yourself and about the things that are important to you to a few close friends, and if this kind of self-disclosure is important to you, then we can say that deep self-disclosure to close friends is part of your interpersonal style and that it is one of your interpersonal values. Or suppose that you never gossip or talk about people behind their backs, because you think that that's a mean thing to do and because you don't want to ruin anybody's reputation or get anybody into trouble. In this case, not gossiping about others is part of your interpersonal style, and it is one of your interpersonal values. As a final example, suppose that, as you look at the ways in which you act with people (your interpersonal behavior), you notice that you usually get your way when you're with your friends. For instance, you get your friends to go to the movie that *you* want to see, or *you* always set the time when the group is going to play basketball, or *you* usually choose the topic of conversation and

11

do most of the talking, or you have a way of getting others to do things for *you*. In this case, controlling or dominating or manipulating others is part of your interpersonal style and one of your interpersonal values. This laboratory group experience will give you a chance to take a good look at your interpersonal values. In fact, you might be surprised to find out what some of them are. Although you'll probably want to keep and strengthen many of your values, it's possible that you'll want to change or get rid of others. However you decide to strengthen or change your values, the group is a place where you can begin.

Values and conflict. A third reason for discussing values here is that conflicts take place if different values get in the way of one another. This kind of conflict can go on either *inside* you or *between* you and another person. Let's look inside you for a moment. If you value both personal comfort and helping others when they make reasonable requests for help, you might sometimes find yourself in conflict. Helping others means that at least at times you have to put your own comfort aside, that you have to put yourself out. However, if being comfortable is a more important value for you than helping others, then you probably won't help others very often at all. In this case, you're in conflict with yourself, because your two values (comfort and helping others) don't fit together very well in practice.

Differences in interpersonal values can also lead to conflicts *between* you and another person. For instance, if you tend to get your way all the time, and if I'm the kind of person who doesn't want to cause any trouble and who always gives in to what others want, then you and I probably won't be in conflict very much. But, while our relationship would be peaceful, it wouldn't be too healthy or growthful for either of us. However, if you and I both highly value getting our own way, we're going to be in constant conflict. Let's take another example. If you are my friend and you value talking deeply about yourself, and if I value keeping my private thoughts and feelings to myself, we might end up in conflict. If you reveal yourself to me—especially if you tell me important things about yourself—you'll probably expect me to do the same. If I don't, you'll probably end up getting angry with me: "I tell you practically everything about myself, but you never tell me what's going on inside you!" This kind of complaint is heard fairly often between friends.

As you talk with your fellow group members, you'll probably discover that your interpersonal values are different from some of theirs, and these differences may lead to conflict: "You never talk to me in this group unless I begin the conversation. I'm beginning to think that you don't really care to talk to me." Conflict is *not* a bad thing *if*—if it's not avoided or ignored and if it's faced and discussed sensibly (by avoiding arguing for the sake of arguing, by refusing to get into name-calling, and by maintaining respect for one another). Research studies show that conflict, if faced sensibly, helps people grow. These studies also show that, if conflict is ignored, it doesn't just go away. It hangs around and comes out in sneaky ways, such as coldness, boredom, and indifference to what is happening in the group. Conflict in interpersonal values gives you and the other people in your group an opportunity to examine your own interpersonal values more carefully. In the group you can see different kinds of interpersonal values in action, you can get other people's viewpoints, and you may even decide to drop some of the values you now have and to add others.

The interpersonal values "pushed" by this book. Finally, it's important to bring up the question of interpersonal values at this point because this book "pushes" certain interpersonal values. For instance, in this book it's assumed that the following goals are "good"—that is, that they represent values everyone could profitably adopt:

- developing a wide variety of interpersonal skills;
- working at trying to understand others, especially those close to you and those who might legitimately ask for your help;
- developing the skill of communicating to others that you understand what is going on inside them;
- being open and able to talk deeply about yourself with others when the relationship calls for it—for instance, close friend with close friend, husband with wife;
- developing the skill of challenging others (especially when those close to you would like to be challenged or when you are in a position that demands that you challenge others).

These and the other interpersonal values "pushed" in this book will become much more concrete and clear to you as you read the book and especially as you move through your group experience.

The purpose of this group experience is not to force you to accept and practice a particular set of interpersonal values. The purpose is rather to have you look at, learn, and experiment with certain interpersonal values and social skills. It is my belief that these skills and values will sell themselves. Moreover, if you have problems with the values pushed by this book, you have every right to push back.

Interpersonal Style: What Is Yours Like?

The term *interpersonal style* has already been used several times in these pages and is going to come up again and again in the pages to come. For instance, in the next chapter I'll explain how finding out what your own interpersonal style is like is one of the most useful goals of the laboratory experience. It's important, then, to get a clear understanding of what the term means. Let's start with a definition or description:

> *Interpersonal style* refers to the usual, the ordinary, the day-to-day ways in which you behave or act when you're with other people (family, friends, people you work with, acquaintances, strangers, people in authority). It includes both the skills you use and the skills you lack, both your successes and your failures in relating to others.

Such a description still isn't concrete or clear enough for our purpose in the group, so I'll try to make it clearer and more concrete. If I were to ask you "What are you like when you are with people, how do you act?" you might answer "Well, it depends on the people I'm with. With friends I'm comfortable—I talk easily, but I don't like to argue. With strangers I'm very uncomfortable. I feel very shy and usually don't know what to say. When I'm with my family, I'm rather quiet. We don't talk a lot to one another." By answering my question in this way, you've given me a short description of your interpersonal style—how you act, what you do, how you think and feel when you are with people. Notice that your answer also indicates that your interpersonal style isn't a simple thing. It's complex, because you act in different ways with different kinds of people. And perhaps you act in one way when you're in a bad mood and in another way when you're in a good mood.

One way to make the term *interpersonal style* come alive is to ask yourself the following questions. As you begin to ask yourself these questions, you'll notice two things. One is that the questions will force you to think about yourself in relationship to other people (maybe you don't usually think about the things these questions ask) and to think about yourself in clear, concrete, and specific ways. The second is that you'll notice how complicated relating to others can be. Put more positively, relating to others isn't just complicated; it's a *rich* experience.

Some Questions about My Interpersonal Style

- *How big a part of my life is my interpersonal life?*
  - How much of my day is spent relating to people?
  - Do I want to spend a lot of time with people, or do I prefer being by myself, or am I somewhere in between?
  - Do I have many friends or very few?
  - Whether many or few, do I usually spend a lot of time with my friends?
  - Is my life too crowded with people?
  - Are there too few people in my life? Do I feel lonely much of the time?
  - Do I prefer smaller gatherings or larger groups? Or do I prefer to be with just one other person most of the time?
  - Do I plan to get together with others, or do I leave getting together to chance—if it happens, it happens?
- *What do I want and what do I need when I spend time with others?*
  - What do I like in other people—that is, what makes me choose them as friends? Is it intelligence or physical attractiveness? Is it the fact that they're good-natured and pleasant or that they have the same values as I do? Do I choose to be with people because they're important or in positions of authority?
  - Do I choose to be with people who will do what I want to do?
  - Do I choose to be with people who will take over and make decisions for the two of us?
  - Do I just spend time with whoever happens to come along?
  - Are the people I go around with like me or different from me? Or are they in some ways like me and in other ways different? How?
  - Do I feel that I need my friends more than they need me, or is it the opposite?

15

- Do I let others know what I want from them? Do I let them know directly, or do they find out what I want in indirect ways?
- *Do I care about the people in my life?*
  - If I care about others, how do I show it?
  - Do others know I care about them?
  - Do I take others for granted?
  - Do I wonder at times whether I care at all?
  - Do I see myself as selfish or generous or somewhere in the middle?
  - Do others see me as self-centered? If so, how?
  - Do others care for me? How do they show it?
- *Am I good at relating to people? What are my interpersonal skills like?*
  - Am I good at both understanding others and letting them know that I understand?
  - Do I respect others? How well do I communicate that I do respect them?
  - Am I my real self when I'm with others, or do I play games and act phony at times?
  - Am I open—that is, willing to talk about myself—when I'm with people who want to be intimate with me?
  - Can I confront others without trying to punish them or to play the game of "I'm-right-and-you're-wrong"?
  - Do I ever talk to others about the strengths and the weaknesses of our relationship?
  - Do I make attempts to meet new people? Does the way in which I meet new people encourage them to make further contact with me?
  - Am I an active listener—that is, do I both listen carefully and then respond to what I've heard?
  - Do people I know come to me when they're in some kind of trouble? If they do, do they leave me feeling understood or helped?
  - Am I outgoing, a go-getter in my relationships, or do I sit back and wait for others to make the first move?
- *Do I want to be very close to some people?*
  - What does closeness or intimacy mean to me? Does it mean deep conversations? Does it mean touching and being physical?
  - Do I enjoy it when others share with me whatever is

important in their lives, including their secrets and their deepest feelings?

- Do I like to share whatever is important in my life with others, including my secrets and my deepest feelings?
- What people am I close to now?
- Do I encourage certain others to get close to me? How do I do it?
- Does closeness frighten me a bit? If so, what is it about closeness that frightens me?
- Are there many different ways of being close to others? What are these ways? Which ways do I prefer?

- *How do I handle my feelings and emotions when I'm with others?*
  - Do others see me as a very feeling person, or do they see me as rather cold and controlled?
  - Which emotions do I express easily to others, and which do I tend to swallow or hide?
  - Is it easy for others to know what I'm feeling?
  - Do I let my emotions take over and rule me when I'm with others?
  - Do I try to control others by my emotions—for instance, by being moody? Do I manipulate others?
  - Do I think that it's all right to be emotional with others?
  - How do I react when others are emotional toward me?
  - Which emotions do I enjoy in others? Which ones do I fear?
  - What do I do when others keep their emotions locked up inside themselves?

- *How do I act when I feel that I'm being rejected by someone?*
  - Does feeling left out and lonely play much of a part in my life?
  - If I feel rejected, how do I try to handle my feelings?
  - Do I sometimes avoid trying to get to know someone or joining a group of people because I'm afraid that I will be rejected?
  - Can other people scare me easily?
  - Have I ever really been let down or rejected by someone?
  - How easily am I hurt, and what do I do when I do get hurt?
  - Do I ignore or reject others who might want to get closer to me?

17

- What do I do when others want to get closer to me and be my friends and I don't want them to?
- *Do I want a lot of give-and-take in my relationships with others?*
  - Do I play games with others, or do I prefer to be straightforward and direct with them? Do people play games with me?
  - Do I like to control others, to get them to do things my way? Do I let others control me? Do I give in to others much of the time?
  - What do I ask of my friends? What do my friends ask of me?
  - Are there ways in which my friendships or my other relationships are one-sided?
  - Am I willing to compromise—that is, to work out with another person what would be best for both of us?
  - Do I think that it's all right for others to influence me and for me to influence others, within reasonable limits?
  - Do I expect to be treated as an equal when I'm with others? Do I want to treat others, especially my friends, as equals?
  - Do I feel responsible for what happens in my relationships with others, or do I just let things "take their course"?
- *How do I get along in my work and school relationships?*
  - How do I relate to people in authority?
  - At school or at work, do I treat people as people, or do I see them as just other workers or just other students?
  - Am I so personal at school or at work that I don't get my work done?
- *What are my main interpersonal values?*
  - Do I want to grow in my interpersonal life and relate better to others?
  - Am I willing to work, to risk myself, to put myself on the line with others in order to get involved in a richer interpersonal life?
  - Am I willing to allow others to be themselves?
  - Is it important for me to be myself when I'm with others?
  - In what ways am I too cautious or too careful in relating to others? What are my fears?
  - Do I get along with people who have opinions and views and ways of acting that are different from mine?

- Do I have any prejudices toward other people?
- How straight or rigid or unbending am I in my relationships with others?
- Do I share my values with others?
- Do I put so much emphasis on interpersonal relationships (for instance, my friendships) that they interfere with my work or with my other involvements in life?

Seeing so many questions about how you relate to other people may make your head spin a little. Still, if you were to answer at least some of the questions in each of the ten sections or categories above, you would begin to form a statement or a description of your own individual interpersonal style. It isn't necessary for you to answer all these questions right now, but it *is* important to get a clear idea of what is meant by the term *interpersonal style.*

Self-Awareness

As you review the preceding questions from time to time, you'll become more and more aware of what goes on inside you as you go about relating to others. Many of the exercises in this book are exercises in *self-awareness*—that is, they encourage you to look inside yourself and to ask such questions as these:

- What do I think about myself?
- How do I feel about myself?
- How much do I care about myself?
- Do I ever try to hide the truth about me from myself? Do I play games with myself?
- In what ways am I good to myself?
- In what ways am I unkind to myself?
- What makes me feel good about myself?
- What makes me get down on myself?
- What are my strengths?
- What are my weaknesses?
- Do I have a reasonable understanding of myself? Do I spend too much time thinking about myself?

Friendship is, in one sense, a gift of yourself to another person. But, if you don't like yourself—that is, if you don't see yourself as

19

much of a "gift"—then it will be difficult for you to establish friend-ships. If you don't even know that you don't like yourself, establishing relationships will be even more difficult. How you feel about yourself, then, is very important, because it influences the ways in which you relate to others. If you're angry and dissatisfied with yourself, you may express your anger by being angry and dissatisfied with others. On the other hand, if you're at peace with yourself and possess rea-sonable self-esteem, then it's more likely that you'll relate more easily to others.

It has been said that self-knowledge is the beginning of wis-dom. It can also be the beginning of better relationships with others. Therefore, developing the skill of self-awareness, especially awareness of yourself in relationship to others, is an important goal of this learn-ing program.

Your Reasons for Being in a
Human-Relations-Training Group

If you ask yourself "Why am I getting myself into this group experience?" what answer do you come up with? There are many possible reasons for your being here. Some of the reasons given by those who participate in group experiences are:

- My friends are doing it.
- I want to find out how others see me.
- It's a required course.
- I want to see whether I have the interpersonal skills the course talks about.
- I feel I'm not very good at relating to people.
- Others have told me that it's a good course.
- I'm curious to see what a group experience is like.
- I'm lonely; I might find some friends here.
- I want to grow as a person; this sounds like it might help.
- I'm too passive; I want to learn how to stick up for my rights.
- I want to learn some basic interpersonal skills.
- The teacher is an attractive person.

Whatever your reasons for coming, you might discover new ones as you actually participate in the training group. But, as we've already noted, you will get little from the group unless you give

20

yourself to the skills training and the group experience freely, without deciding beforehand that it won't be worthwhile, that it will be too much work, that it will be too artificial, and so forth. Therefore, at this time, you're asked to enter into a *contract*; that is, you're asked to give yourself freely to the training program as it is outlined in this book. The next chapter is a summary or an overview of the group experience, including both the goals of the laboratory and the training experiences that will be used to reach these goals. Both goals and training procedures will be discussed in terms of a contract.

Exercises in Self-Awareness

The following exercises ask you to reflect on yourself in ways that you may not be used to. Your instructor may discuss these exercises with you. Try to get a clear idea of what you want to achieve in these exercises before you actually do them. Basically, they're meant to help you develop the skill of self-awareness.

Exercise 1
Reviewing the Questions on Interpersonal Style

a. From the list of questions about your interpersonal style that I gave earlier, pick out the ones that you would like to answer for yourself. Which questions capture your interest the most? Try to answer at least a couple of questions in each section or category.

b. Which questions do you find most difficult to answer? Which questions would you be afraid to discuss in your training group?

Exercise 2
This Is Me

Using the set of questions as a guide, write a short, one-page description of your present interpersonal style. On the one hand, write only what you would be willing to read or show to your fellow group members. On the other hand, try to show a side of you that you think or feel you don't usually let others see.

The Contract:
An Overview of
the Training Process

In this chapter I outline the entire training experience in the form of a contract. Obviously no one can force anyone else to learn and practice interpersonal-communication skills. However, I do suggest that the person who makes a commitment to learning these skills (that is, the person who "contracts" to learn) will be the most effective learner. Therefore, whether or not your instructor asks you to enter into a formal learning contract isn't the point. What is important is the kind of learning contract you make with yourself.

Introduction

Contract and discipline. A piano player needs a great deal of *discipline* in order to become a good pianist. Discipline means, at least in part, self-control. A person is disciplined if he or she makes whatever sacrifice is necessary in order to achieve a goal. Thus, discipline often involves some kind of hardship—doing things that aren't pleasant and giving up things that are. A pianist, for instance, has to practice hard every day, even when he or she doesn't feel like practicing.

Some people say that they don't like discipline because they want to be spontaneous and free. But who is freer, the person with discipline in his or her life or the person without? I believe that it's impossible to be spontaneous or free without discipline. A person who has discipline is the master or mistress of his or her life. He or she has both the time and the energy to be spontaneous. For instance, a college student who studies regularly instead of waiting until the day before a test to "cram" can decide to go on a trip the weekend before the exam. In this case discipline doesn't cut down on freedom or spontaneity; rather, it increases it. Let's take another example. A basketball player who spends a great deal of time and energy practicing many different moves on the basketball court and who is in good enough physical shape to make these moves is a person with discipline, at least in the area of basketball playing. In a game such a player can be much more spontaneous than a player who has decided that only a few moves are necessary. Because the second player lacks the discipline needed to learn the appropriate skills, he or she cannot be nearly as spontaneous on the court.

Learning interpersonal skills takes hard work and discipline. In this book you are asked to enter into a contract, to give yourself to the step-by-step process of learning these skills. Such systematic training is itself a form of discipline. In this chapter I will try to give you as clear a picture as possible of what is going to be asked of you. The rest of the chapter, then, is both an overview of the training program and a description of the contract itself.

Clear contracts. Trouble arises when contracts between people aren't clear or when one person makes a one-sided contract that he or she hides from the other party. A hidden contract doesn't have to be intentionally deceptive. It's simply one that isn't shared with the other party. For instance, suppose that two people, whether deliberately or not, each make a hidden or unspoken contract when they get married:

> *He* (to himself): "In our marriage my wife will be one of my best friends, but I will continue to have other very good friends, both male and female. I expect to spend time with these people, sometimes together with my wife and sometimes alone."
>
> *She* (to herself): "My husband will be my one really true friend. We'll spend time with other people, but we won't be that close to others. That is, we won't share with them the deep things that we share with each other."

Few people make hidden contracts this clearly and knowingly, but, knowingly or not, hidden contracts *are* made often enough, and they interfere with interpersonal relationships.

Hidden contracts can also cause trouble in human-relations-training groups. Consider the following hidden, one-sided contracts:

Group participant A (to self): "I'm going to play it cool in this group. I'm not going to take any risks. I'll do as little as I have to do to get a passing grade."

Group participant B (to self): "I've heard that the only way to survive in these groups is to attack. I'll be brutally open and frank and honest. I'll tell people what I think of them right away. That will make them afraid of me and keep them off my back."

Group participant C (to self): "I'm a lonely person. I want to be closer to people. So I'll have to be very careful how I treat others in the group. I won't challenge them, because they might reject me if I do. I want to make some friends, not enemies."

In each of these cases, a group member has decided how he or she will act in the group without any kind of consultation with any of the other members of the group. Again, few people make these kinds of hidden contracts so knowingly or deliberately, but, deliberately or not, such contracts are sometimes made in human-relations-training groups, and they do get in the way of learning in such groups. Once the members of the group agree to the following contract, it's much less likely that hidden contracts will be made. If they *are* made, however, they can be challenged more easily.

Contract Overview

Let's first see in outline form the kinds of learning experiences the contract will ask of you.

I.   *Learning different kinds of communication skills.* There are four sets of basic communication skills that you will be asked to learn.
  A.   *The skills of making yourself known to others*
   1.   *Self-disclosure:* the skill of talking about yourself appropriately to others (Chapter 3)
   2.   *Feelings:* the skill of talking about and expressing your feelings and emotions (Chapter 4)
   3.   *Concreteness:* the skill of speaking clearly and specifically about yourself, your feelings, your experiences, and your behavior (Chapter 5)

25

B. *The skills of responding to others*
1. *Attending:* the skill of paying attention to others and listening to them in an active way (Chapter 6)
2. *Understanding:* the skill of communicating to others an understanding of their feelings, experiences, and behaviors (Chapter 7)
3. *Respect:* the skill of communicating respect for another person through your interpersonal behavior (Chapter 8)
4. *Genuineness:* the skill of communicating genuineness through your interpersonal behavior (Chapter 8)
C. *The skills of challenging others*
1. *Deeper understanding:* the skill of helping others see themselves as others see them (Chapter 9)
2. *Strengths:* the skill of being able to recognize and point out to others their good points and their strengths (Chapter 10)
3. *Confrontation:* the skill of inviting others to examine and possibly change their behavior (Chapter 10)
4. *Immediacy:* the skill of being able to talk to another person about how your relationship with him or her is going (Chapter 11)
D. *Group communication skills* (Chapter 12)
   These are the skills you need in order to be an active, participating member of a group. Being part of a group is a demanding communication experience that calls for a number of different skills. Some of the skills discussed are:
   • the skill of being active rather than passive in a group;
   • the skill of involving yourself in the conversation of others;
   • the skill of challenging the members of a group;
   • the skill of getting in touch with the mood of a group.

II. *Establishing and developing relationships with the other members of your group.* As you learn the skills listed above, you will be asked to use them to try to establish a relationship with each of the other members of your group. Later on in this chapter the meaning of establishing and developing relationships will be discussed at greater length.

III. *Exploring your interpersonal style.* The meaning of the term *interpersonal style* was discussed in Chapter 1. You'll be asked to

examine and learn about your own interpersonal style by watching yourself in action—that is, by:
- watching how you involve yourself with your fellow group members as you learn the skills listed above;
- watching how you go about trying to establish and develop some kind of relationship with each of your fellow group members;
- watching how you usually relate to different kinds of people outside the group;
- getting feedback from your fellow group members on your interpersonal skills, how you use them in the group, and what you're like when you relate to different people in the group;
- watching how you affect people both inside and outside the group.

All of these ways of exploring your interpersonal style are also ways of helping you develop the skill of self-awareness.

IV. *Trying to develop and change your interpersonal style* (Chapter 14). The fourth and final task has to do with changing your interpersonal behavior. If you examine your interpersonal style well, you'll probably find some things you do that you would like to stop doing and some things that you don't do that you'd like to start doing. For instance, you might discover that you confront or challenge others very harshly and that, as a result, they get turned off and don't listen to what you're saying. Although you wouldn't want to eliminate confrontation from your interactions with others, you might well want to eliminate the harshness of your approach.

These, then, are the four goals or tasks that constitute the contract for this training experience. In practice, some of the steps in the outline will overlap. The way of going about these four tasks will be specified by your instructor.

Now that we have an overview of the four goals or tasks of this training program, let's look at each one more closely.

Task I: Learning Different Kinds of Communication Skills

After reading the preceding list of skills, you probably thought to yourself "I think I already have some of these skills." No doubt you do. However, in this training program it won't be assumed that you either do or do not possess any specific skill. You'll be trained in each skill. One way of looking at the skills-training program is as an opportunity to get an interpersonal-skills checkup. If you find that you do have a particular skill, then you can use the program to strengthen that skill. For instance, you might discover that you're a fairly good listener. You pay attention to others, and you try to communicate to them that you understand what they're saying. In this case, you would merely try to improve a skill that you already possess. However, you might also discover that you never talk very much about yourself, because you find that difficult to do. You don't know what to say, or you feel embarrassed when you talk about yourself, or you're afraid to share your feelings with others. In this case, you can use the training program to develop a skill you feel you don't have.

Task II: Establishing and Developing Relationships

In Chapter 1 the expression *interpersonal style* was clarified by means of a number of questions for you to ask yourself concerning just how you go about relating to other people. Here it will be helpful to clarify what *establishing and developing relationships* means. Establishing (starting up) and developing (working at) relationships includes such things as:

- *spending time with another person.* You'll spend some time with your fellow group members, even though the time will be limited.
- *doing things with this person.* In the training group you'll work with others in developing communication skills. You'll be talking about your own interpersonal style and listening to others talking about theirs.
- *developing a respect and perhaps even a liking for this person.* The relationships you establish with your fellow group members need not be all alike. Obviously you may come to like some more than others. However, respect, which means being "for" another just because he or she is a human being, seems to be owed to every group member. Seeing how differently your relationships with others develop can be one of the most interesting parts of the training lab.

- *caring about, being concerned about, this person.* In the training group, caring means that you want to cooperate with your fellow participants in pursuing the goals of the group. It means that you'll try to help them to learn the skills and to explore their interpersonal styles and that you'll expect them to do the same for you.
- *feeling at home with this person.* Feeling at home doesn't mean that you and your fellow group member become so comfortable with one another that you never challenge one another. But developing a relationship does mean that you're not always on edge in the presence of the other person. People usually feel at home with others when they feel respected and accepted.
- *developing a sense of give-and-take so that the relationship is not one-sided.* A relationship goes both ways. If you merely talk about yourself to another person without that person's sharing himself or herself with you, you don't really have a relationship. In a relationship you both give and receive.
- *becoming willing to talk about personal issues with this person.* Since the group deals with interpersonal style, developing a relationship necessarily means talking about yourself personally and expecting that others will do the same. A relationship is always some kind of sharing. Here you share yourself by talking about yourself.
- *giving and getting feedback.* In a work relationship you expect to get feedback on how you're doing from your supervisor; that's part of the relationship. In the training group an important part of the relationship is giving and getting feedback on skills development and interpersonal style.
- *being willing to both give and get help.* Giving and getting help doesn't mean that you're in the group in order to be someone's counselor or someone's client. Part of friendship is being able to give help without becoming a counselor and to receive help without having to become a client.

These are some of the behaviors that are involved in establishing and developing relationships in the training group. Perhaps you can think of others. What are your expectations of the relationships that you're about to form in the training group?

The question of artificiality. Obviously, establishing and developing relationships in the group won't be exactly the same as doing these things naturally, in your everyday life. Most of the time relationships just seem to "happen." However, even in everyday life there are times when you decide that you want to get to know someone and set out deliberately to establish a relationship. You do so, ordinarily, because you're attracted to the person in one way or another and therefore want to get closer to him or her. In this laboratory group, however, you're asked to form and develop relationships with a particular set of people merely because they are members of the group. Yet they're people like you—that is, people who are interested in improving the ways in which they relate to others. Therefore, even though these relationships may last only for a short time, they are real and can be very helpful in pursuing the goals of the group. Saying that the group is, in a sense, artificial is in no way the same as saying that the relationships you establish are fake or phony.

Investing yourself in these relationships. Just as you can speak of investing money in a business in order to make a profit, you can speak of investing *yourself* in the members of the group in order to profit by growing interpersonally. If you invest your money in a savings-and-loan institution, both you and the institution make money. In the same way, if you invest yourself in your fellow members, both you and they can profit.

Thinking of a relationship as an investment of yourself—that is, as a commitment to another person—is another way of clarifying what "establishing and developing a relationship" means. In this group you're asked to invest yourself in or commit yourself to your fellow group members for the duration of the group. This investment includes:

- working cooperatively with others in trying to achieve the goals of the group;
- learning how to listen to others in an active rather than in a passive way—listening because you *want* to listen;
- working at understanding others—trying to see the world as they see it;
- gradually letting others inside your world, inside you—talking about yourself in deep rather than superficial ways;
- letting others know what you like about them, what you see them doing well;
- sharing with them what you learn about yourself during the course of this group experience;

- letting others know what holds you back from getting involved with them, what they do that scares you or makes you annoyed or makes you want to move back;
- asking others for feedback on how well you're doing in the group and what you might do to improve your skills and your interpersonal style;
- talking about the feelings and emotions that you experience because of what is happening in the group.

I'm sure that you can think of other ways of investing yourself in the group. Establishing and developing relationships is a very active process, one that takes a great deal of work. Some of us don't relate better simply because we don't want to expend the energy it would take to do so. While we like the idea of relating more effectively to others, we like our comfort more. However, a boring and sterile interpersonal life can be the price of too much interpersonal comfort. One way of looking at this laboratory in interpersonal learning is to see it as a course in responsible and reasonable *risk taking*.

Task III: Exploring Your Interpersonal Style

Task III is closely related to Tasks I and II. In the group you find out what your interpersonal style is like by actually involving yourself with people. In order to learn communication skills, you communicate with others. Then you begin to use these skills in establishing and developing relationships. All of this involves your interpersonal style. As you go about Tasks I and II, you both watch yourself and get feedback from others. For instance, as you watch yourself learning interpersonal-communication skills (Task I), you might say to yourself:

> "I notice that I'm trying to learn these skills better than anyone else in the group. A need to be the best—that seems to be a significant part of my interpersonal style."

Once you discover how competitive you are, you can share that with the other members of your group. You can also decide whether that's the way you want to be or whether you would rather tone down your competitiveness a bit. Or, as you try to talk concretely about yourself, you might say to yourself:

"I didn't realize how private a person I am. I'm finding it very hard to talk about myself and my feelings in very concrete terms. I'm very careful about whom I let inside me."

So, even as you're learning communication skills, you're also finding out a lot about yourself. Sharing what you learn with the other members of the group is one way of exploring your interpersonal style.

Task II, establishing and developing relationships, also provides you with excellent opportunities for exploring your interpersonal style, especially if there are different kinds of people in your group. For instance, you might find yourself saying something like this in your group:

"Don, I find it difficult trying to begin a conversation with you. You're relatively quiet in the group. That means I have to put out a lot of effort if I want to get closer to you. What I'm learning is that I'm a bit lazy in my interpersonal life. If getting to know someone is not an easy task, I tend to forget it."

Here, trying to establish a relationship with Don has taught you something about your general interpersonal style. Or something like this might happen:

"Kathy, I don't actually feel like talking to you today. Last week, at least in my way of looking at things, you practically told George that you didn't want anything to do with him. And you did it in a kind of harsh, 'that's-it' way. I learned something about myself. I practically never risk trying to get close to someone unless I more or less know ahead of time that that person is going to accept me. What happened between you and George makes me afraid that you might not accept me. And so it's very hard for me to talk to you right now."

Watching how you react to the other members of the group and then discussing it in the group is one way of exploring your interpersonal style. You get to know new things about yourself, or you see more clearly what you already know.

Getting feedback from your fellow group members as you go about the tasks of learning skills and establishing relationships is also

32

an excellent way of getting to know more about your interpersonal style. For instance, someone might say to you:

> "I think that you're a very perceptive person. The insights you offer are clear and concrete. And you give feedback in a very caring way. But you always sound like you're apologizing when you give feedback. I sometimes get the feeling that you don't think much of yourself because you're always putting yourself down. That makes it harder for me to listen to the good things you have to say."

As you learn about yourself both by watching yourself interact with others and by getting feedback, you can begin to put together a picture of what you're like in interpersonal situations. Then you can begin to think about what you would like to change in your interpersonal style.

Task IV: Trying to Develop and Change Your Interpersonal Style

The training group is called a "laboratory" in order to encourage you to experiment with what you're learning. In learning new forms of behavior, it's useful to have a place where you can safely try out these behaviors until you feel comfortable with them. For instance, once you learn an interpersonal skill such as the communication of understanding, you can experiment with it by using it in your conversations in the group. You can also experiment with the way you go about establishing and developing relationships with others (Task II). For instance, if your ordinary style is to talk about yourself and share things that are important to you without expecting the other person to do the same, you can experiment with a different approach by *asking* others to share themselves just as you share yourself.

Experimenting with new interpersonal behaviors is part of the process of changing your interpersonal style. The process of change may involve something like the following sequence.

1. *Skills*. Learning new communication skills or improving the skills you already have is in itself a way of changing your interpersonal style. For instance, taking the time to listen to others more carefully before responding to them is a different way of being or communicating with others.

2. *Self-awareness*. As you become more and more aware of what your interpersonal style is like by consciously developing your own self-awareness skills and by getting feedback from others, you'll soon realize what you want to strengthen and what you want to improve in your interpersonal style. For instance, if you begin to realize that you let others influence you too much, you can decide that you're going to learn how to make your own decisions.

3. *New possibilities*. Once you realize what some aspect of your interpersonal style *is* like (self-awareness), you can begin to get a picture of what you *want* to be like. You can form such a picture by reading books like this one, by watching others with styles different from your own, or just by getting new ideas of what you want to be like as you learn and practice communication skills and try to develop relationships with the other members of the group. For instance, you observe that, when one of your fellow group members receives any kind of criticism, she makes an effort to understand as clearly as possible what the person who is offering the criticism is saying before responding. You like this approach—you see it as a new way to respond to criticism. Here, what you've observed another person doing becomes a new possibility for your own interpersonal style.

4. *Experimenting in the group*. Your next step is to try out this new interpersonal behavior in the group. For instance, when you receive criticism in the group, you try to respond without being defensive. You use your skill of understanding to make sure that you know exactly what the other person is saying before you respond to it.

5. *Changing outside the group*. Once you see that some new interpersonal behavior works for you inside the group, then you can begin using it outside the group—perhaps slowly at first, until you feel comfortable with the change.

The Work and the Rewards

You'll probably find that some skills come more easily than others. For instance, you might find it rather easy to communicate to others that you understand them but much harder to challenge or confront them. It takes time and patience to develop a well-rounded or complete set of interpersonal skills. The task will be much easier if you and your fellow group members help one another to learn. Some members of the group will learn these skills more quickly than others. A real sign that you possess a skill fully, however, is your helping someone else who is having a harder time. One of the most rewarding experiences possible in a human-relations-training lab is the development of the group into a learning community in which the members care for one another, trust one another, and make demands for growth on one another.

There's a lot to do in your group and relatively little time to do it in. If you're in a group that has eight members, each member has seven different relationships to establish and develop. Altogether, in a group of eight people there are 28 separate pairings or relationships. Since that adds up to more work than can be finished in the time allowed to the group, there will be unfinished business at the end of the course. But the group experience is a preparation for more effective involvement in everyday life. The training group is a beginning; the real payoff is in your day-to-day interpersonal relationships.

I've said that pursuing the four tasks outlined in this contract is a lot of work, and it is. However, learning skills and establishing new relationships can also be both exciting and fun. It's very rewarding to learn skills and to use them to establish solid relationships with others, both in the group and outside.

Small-group learning. This book assumes that much of your learning will be experiential and that it will take place in a small group, but this may or may not be the case. If you're going to use the small group as a means of learning, it would be helpful to read Chapters 12 and 13 early, since these chapters deal with the skills needed to communicate effectively in a small group. Chapter 14 deals with changing parts of your interpersonal style. Since experimenting with change is part of this laboratory, you might want to read Chapter 14 early in the course also.

Exercise 3
A Log Leading to an Agenda

This exercise is to be repeated between meetings. A log is a written account of the important things that happen between you and others in your group meeting. An agenda is a plan for what you would like to accomplish in the next meeting.

The Log

After each group meeting, write down the things that were for you the highlights or most important happenings in that meeting. Also write down what you feel *now* about what happened in the meeting. Here are some examples of what might appear in your log.

- "I didn't do well in practicing the skill of understanding. I feel bad about myself right now. Others are doing better than I am. I usually feel pretty bad when others do better than I do."
- "Jean told me that I am warm and understanding. That flatters me a lot, especially since I'm attracted to her."
- "Even though I now possess the skill of understanding, I don't use it very much in the group. My tendency is to confront others without first letting them know that I understand them. People are beginning to turn me off. I feel isolated, and I'm not sure how to get back into the group."

The log should be as concrete as possible. It can include *experiences* (what happens to you in the group, what others do to you, how they react to you), *behaviors* (what you do concretely and specifically in the group), and *feelings* (how you feel about yourself, how you feel about others and what they do). Your log should include where you stand with each of the other members of the group from week to week. For instance, it may include such statements as:

- "Jack and I still don't talk to each other. It's like a game, each wondering when the other one is going to break the ice. I never realized that I'm this stubborn."
- "I'm very attracted to Peter. I spend a lot of time talking to him. I think he likes me a lot, too. The others must notice this. I wonder how they feel when they see that I'm more interested in Peter than in them."
- "Cindy is a very active group member, and she talks to me a lot, but I still don't know what she thinks of me. She doesn't seem to commit herself to anyone, certainly not to me. I find it hard challenging her on this, since she's so active. She's very good at understanding others, but she doesn't reveal very much about herself."

Forming and developing relationships with each of the other members of the group is one of your most important activities, and the log is a way of keeping track of what is happening in these relationships.

The Agenda

Writing the log isn't like keeping a diary; it's not a goal in itself. Your log has the very practical purpose of helping you plan what you would like to do and accomplish *in the next meeting.* Your plan of action is called the *agenda.* For instance, suppose you write this in your log:

- "I don't talk to Betty at all, because I think she doesn't like me or at least is indifferent to me, while I'm attracted to her. I don't like this combination."

In this case your agenda might read:

- "Talk to Betty. Tell her your feelings. Clear the air. It's no use just avoiding her, and you must admit that you don't really know how she feels. It seems that you'll just have to risk getting a little hurt."

Concrete logs lead to concrete agendas. I'm not suggesting that it's easy to put concrete agendas into practice. I *am* suggesting, however, that concrete agendas make it more *likely* that you will act. The purpose of the agenda is to make sure that everyone comes to the group meeting prepared. Being prepared avoids wasting time asking yourselves "Well, what shall we do today?" Sometimes people write very good logs and draw up very good agendas but then fail to put their agendas into practice. If this is your problem, at least include it in your next agenda:

"I'm afraid to put my agenda items into practice because I get cold feet. Next meeting I'll tell the group that I have several logs filled with great agendas and ask them to help me to get to them."

The Skill of Self-Awareness and the Skills of Letting Yourself Be Known

Chapters 3, 4, and 5 discuss how to talk about yourself with others and how to express what you feel about yourself and others. But you can't talk about what's going on inside you unless you're first in touch with whatever that is. For instance, it's impossible to let others know that you're angry if you yourself don't realize that you're angry. It's impossible to let others know the ways in which you like yourself if you never reflect on what you like about yourself. Therefore, it's impossible to discuss the skills involved in self-disclosure and the expression of feeling without discussing the more basic skill of self-awareness. Furthermore, if you're a self-aware person, it's more likely that you'll develop the awareness or perceptiveness needed to understand others. Awareness of self and awareness of others are the foundation of every interpersonal-communication skill.

Before beginning to discuss specific skills, let's take a look at the three essential parts of any interpersonal skill.

The Three Parts of an Interpersonal-Communication Skill

Each of the interpersonal skills that you're about to learn has three essential parts: (1) awareness, (2) communication know-how, and (3) assertiveness. Successful use of a skill means putting all three parts together.

Awareness or perception. We say that a person has awareness or good perception if he or she knows what is going on. A perceptive person is one who pays close attention to people and how they act, thus picking up things that other people don't notice. For instance, a friend of yours says that he's not worried about tomorrow's exam. But he's more silent than usual, he fidgets with his hands, and when he talks, he talks about how he doesn't like this particular teacher. These are all signs that he *is* worried about the exam but for some reason can't admit that to you or perhaps even to himself. You're being perceptive if you pick up cues that tell you what's really going on inside him.

As I've already mentioned, in the skill of letting yourself be known to others, self-awareness is most important. You can't let others know what's going on deeper inside you unless *you* know. Suppose that you tell a boyfriend of yours that you don't care whether he goes out with other girls. He's free, and that's his privilege. However, on evenings that he doesn't call, you get restless. You find yourself casually asking others whether they know whom he's seeing. You're uncomfortable with him when you haven't seem him for a few days and find it hard to talk. If you read these signs correctly, you soon realize that you *do* care that he's going out with other girls. Once you're aware of what's going on inside yourself, you might tell him that it does bother you when he goes out with other girls but that you don't want that to stop him, because you also want the freedom to go out with other boys. You can't tell him this, however, unless you're in touch with your own feelings. Often enough we block or deny our own feelings, which makes it impossible for us to communicate our real feelings to others.

There is an awareness part to every interpersonal skill. Learning how to pay closer attention to yourself, to others, and to the situations in which you find yourself relating to others is necessary if you're going to develop the kind of awareness that is the foundation of interpersonal skills.

Communication know-how. Your awareness or perception of yourself or another person can only be put to use if you know how to communicate this awareness to others. It might happen, for example, that you learn how to pay close attention to others, so that you begin to notice what's going on inside them (you can read the little hints people give that say "I'm angry" or "I'm hurt" or "I'm sad" or "I don't think very highly of myself") but that you don't know how to let them know that you understand. Communication know-how, then, is another essential part of each interpersonal skill. Your accurate perceptions are useless if they remain locked up inside you because you don't know how to express them. Or if the way you communicate your otherwise accurate perceptions confuses or annoys the person you're talking with, you again get nowhere. If you challenge someone in such a way that the person feels that he or she is being punished by you, he or she will probably turn you off, no matter how accurate your perceptions are. On the other hand, if your communication know-how is good but your perceptions are poor, then people will turn you off again. They will see you as a smooth talker with nothing to say. Skillful interpersonal communication means that good perceptions must be joined with good communication know-how.

Assertiveness. Being assertive means actually putting the skills you have into practice. An assertive person steers a middle course between not doing anything and doing too much. An assertive person gets the work of interpersonal communication done but does so in a way that respects both his or her own rights and the rights of others. An assertive person is one who acts, who puts himself or herself on the line, but does so in a responsible way. High-quality awareness and excellent communication know-how are meaningless unless you have the courage to use them. In interpersonal communication, it's the delivery of the message, together with *how* you deliver it, that counts. So an interpersonal-skills-training program is also an assertiveness-training program.

On the other hand, assertiveness without good perception or without communication know-how can lead only to disaster. Perhaps you've experienced the sort of person who, without being very perceptive or having the communication know-how, still is assertive enough to let you know what you're doing wrong. First of all, because such a person lacks perception or awareness, he or she isn't accurate in what he or she is saying to you. Second, the person accuses you or yells at you or scolds you instead of helping you to take a look at your behavior and its consequences. Raw assertiveness without awareness and without communication know-how turns a person into a potentially destructive communicator.

Nonassertiveness and aggressiveness. A person who doesn't speak up in interpersonal situations, who lets himself or herself be manipulated by others, who in communicating to others doesn't get his or her legitimate needs taken care of (for instance, always listening to others but never getting anyone to listen back), who lets others walk all over him or her—such a person is called *nonassertive* or *compliant*. If in some ways or in some situations you are nonassertive or compliant, this lab experience can provide you with an opportunity to explore your nonassertiveness and to do something about it.

A person who gets his or her needs met but in doing so steps on the rights of others is called *aggressive*. For instance, if a waitress fails to bring cream for your coffee and you sit there waiting for her to notice it, you're being compliant. If you call her over, yell at her, and tell her that she's stupid for not bringing it in the first place, you're being aggressive. However, if you call her over and tell her what you need, you're being assertive. This laboratory experience will help you to steer a course between being compliant and being aggressive in interpersonal situations. In my experience with human-relations-training groups, I find more nonassertive than aggressive people.

Exercise 4
How Assertive Am I?

Review your recent interactions with others. Try to discover:

- *one time that you were nonassertive or compliant.* Example: "Last week my brother took three dollars out of my wallet without asking me whether he could borrow it or not. As a result I couldn't go to a movie I wanted to see Friday night. He does things like this, but usually I don't say anything to him."

- *one time that you were aggressive.* Example: "A friend of mine was kidding me about something the other day at school. I had a headache. So I yelled at her and told her that she was an inconsiderate person. I called her immature and then left her standing there."
- *one time that you were assertive.* Example: "I picked up a hitchhiker the other day. He lit up a cigarette. I told him that smoke in such a small enclosed space bothers me, and I asked him to please not smoke while we were in the car. He put out the cigarette."

What are you usually like when you are with others—nonassertive? aggressive? assertive? Are there some situations in which you are usually nonassertive? Are there some interpersonal situations in which you are usually aggressive?

Review the nonassertive example that you've written up for this exercise. What could you have said or done in order to be assertive?

Review the aggressive example you've written up for this exercise. What could you have said or done in order to be assertive rather than aggressive?

Deepening your awareness of yourself and of others and learning the communication know-how that goes with each of the interpersonal skills you are about to learn will help you become more assertive. Becoming assertive in interpersonal situations is an important goal of this laboratory experience.

Self-Disclosure: How to Talk about Yourself to Others

3

This chapter discusses how to talk about yourself in interpersonal situations without saying too much or too little about yourself. It talks about the advantages and the disadvantages of letting others know you better and suggests ways of making what you disclose about yourself both substantial and appropriate. The chapter ends with a few exercises designed to help you decide what you want to tell others about yourself.

Self-Disclosure: Too Much or Too Little?

The issue of self-disclosure, or what we should say about ourselves to others and how we should say it, is a highly emotional one. There are those who believe that people are only too willing to pour out their souls to other people, even to absolute strangers. Such people, the complaint goes, are making pests or nuisances of themselves. Too much self-disclosure is also seen by some people as part of a wider problem in society. People are too willing to undress their souls in public, just as they are too willing to undress their bodies in public (in movies and plays, or on beaches). According to this view, people have lost a sense of privacy; not only are too many people willing to confess their own secrets, but they want everyone else to do so as well. Such public confession, the objectors say, is *not* good for the soul. Rather, all this disclosure is a sign that both individuals and society have abandoned proper controls. Perhaps we could express this complaint in another way by saying that some people are *aggressive* self-disclosers: they tell you about themselves whether you want to hear it or not.

Others admit that a certain amount of excessive nudity, both physical and psychological, is indeed a part of our culture, but they

43

claim that a relatively small percentage of people go to excess. The real problem for most of us, they say, is that we don't reveal ourselves *enough*, even when self-disclosure is legitimately called for. We're too private; we keep too much of ourselves to ourselves; indeed, we don't know *how* to reveal ourselves to others. These people claim that many relationships become boring because people don't know how to share what is going on inside themselves. People fear self-disclosure because they fear getting close to others. In fact, it's said that too little self-disclosure is either itself a psychological problem or at least a sign of a psychological problem. Those who hold this point of view often praise younger people for being more open and honest than the older generation. We could say that, on this view, most people are too *nonassertive* when it comes to self-disclosure.

Obviously, any discussion of self-disclosure is greatly influenced by the values of those doing the discussing. People in the two seemingly opposed camps discussed above seem to have quite different values with respect to self-disclosure. While some people value openness very highly, others value privacy just as highly. Since you will be asked to experiment with self-disclosure, you can use this laboratory as an opportunity to examine your own values with respect to self-disclosure. In this chapter I try to steer a middle course between overemphasizing self-disclosure and underemphasizing it. Perhaps we should call the kind of self-disclosure I have in mind *assertive* self-disclosure. However, what stand you wish to take on your own self-disclosure is, in the final say, up to you.

Overdisclosers and underdisclosers. One way of looking at self-disclosure is to take a look at the two extremes. Some people are overdisclosers; that is, either they talk too much about themselves (the quantity is too much) or they talk too personally about themselves in social situations that do not call for such personal talk (the quality is too much). If, in giving a report in class on some aspect of the American Revolution, you were to talk about the problems you were having with your father at home, you would probably be seen as an overdiscloser. A classroom report is ordinarily not an appropriate occasion for discussing your personal problems. It is perhaps easier to identify an overdiscloser than an underdiscloser, since overdisclosing is more dramatic. Underdisclosers don't want others to know them deeply and therefore speak little about themselves. In fact, they don't speak personally about themselves even when the situation calls for it. For instance, an underdiscloser may encourage friends to talk quite personally about themselves but then refuse to talk personally about himself or herself. Married people frequently complain that they don't know what their spouses are thinking or feeling or even doing. Since marriage is one interpersonal situation in which speaking personally about oneself seems to belong, the husband or wife who says little or nothing about himself or herself could be considered an underdiscloser.

Appropriate self-disclosure. I will use the term *appropriate self-disclosure* to describe the kind of self-revelation that avoids the two extremes of underdisclosing and overdisclosing. By *appropriate* I mean fitting, suitable, the right amount at the right time. First I'll provide a little background to help you determine whether you're an "appropriate" self-discloser. Then I'll explain what I think appropriate self-disclosure means in a human-relations-training group.

Self-Awareness and Self-Disclosure

Let's begin by looking at what some well-known psychologists have thought about the importance of self-disclosure in everyday life and at some of the main reasons why we're afraid to disclose ourselves more fully even to the people who are important to us. This background material does not provide simple answers to the question of self-disclosure. Its purpose is rather to stimulate your thinking so that you can ask yourself questions about the role self-disclosure should play in your life. The values that you would like to hold with respect to self-disclosure may not be entirely clear to you right now; perhaps what is said here will help you to clarify these values.

Underdisclosing and damage to psychological health. A number of psychologists claim that being an underdiscloser can block interpersonal and psychological growth and even lead, in certain cases, to emotional problems. Although they don't have absolute proof for these claims, they do present enough evidence for their case to make us sit up and take notice.

One psychologist, O. H. Mowrer, admits that there are many different reasons why people get into psychological or emotional trouble but suggests that one reason is related to a lack of self-disclosure. According to Mowrer, the first step is taken when people do things that they think they shouldn't do. For example, suppose that a man steals things from the store in which he works, believing that what he's doing is wrong. He might then try to pretend to himself

45

that he doesn't feel guilty. He puts the whole thing out of his mind, or at least he tries to. He thus becomes an underdiscloser to himself; that is, he won't even talk to himself about what he's doing. Next, even though he begins feeling uneasy or nervous—perhaps even getting headaches or pains in his stomach—he still refuses to think about it, much less tell anybody else. He simply rejects the possibility of talking the whole thing over with someone who would understand. He gets more and more nervous (and his aches and pains get worse) until finally he's convinced that he has some emotional problem, though he doesn't know where it's coming from. The solution to this man's problem, Mowrer says, is first to admit *to himself* that he has been doing wrong (confess to himself), then to talk the whole thing over with someone he trusts (confess to someone else), and finally to find some way to undo the wrong he has been doing (give back what he has stolen). In this case, lack of self-disclosure gets the person into psychological trouble, and self-disclosure, together with a change in behavior (returning the stolen goods), is the solution.

Perhaps something similar has happened to you: guilt that you were afraid to admit to yourself or to anyone else got into your guts and began to hurt. At any rate, that's one way that lack of self-disclosure may lead to psychological problems. In this case, confession *is* good for the soul.

Another psychologist, the late Sidney Jourard, held a similar view of self-disclosure (or rather the lack of self-disclosure), but he looked at it from a wider perspective. He claimed that a lack of reasonable self-disclosure can get a person into psychological trouble even when there is no question of wrongdoing. According to Jourard, self-disclosure is absolutely necessary for psychological health and growth. People can't be themselves unless they know themselves. But one of the best ways in which people can get to know themselves most deeply or fully is by sharing their inner selves with another person. People with emotional problems, Jourard went on to say, are often people who work hard at avoiding being known by others. Reasonable self-disclosure is thus both a cause and a sign of a healthy personality. A psychologically healthy person has the ability to make himself or herself more or less completely known to at least one other human being. Furthermore, since love is a gift of self, a person who finds self-disclosure impossible can't really love another human being. People stop growing psychologically, Jourard claimed, when they stop making themselves known to the important others in their lives.

Again, this psychologist did not absolutely *prove* that reasonable self-disclosure is necessary for a full and healthy interpersonal life, but his views should make us think about how we want to share ourselves, especially with the important people in our lives. And, in a sense, the people in your training group will become, for a limited time, important people in your life.

Reasons why we don't share ourselves more deeply with others. There are a number of reasons why we don't share ourselves more fully with others, even important others whom we trust. Most of these reasons are based on some kind of fear.

• *Family background.* How personally do the members of your family talk to one another? Many of your present attitudes toward self-disclosure were probably learned at home. Whether you're willing to talk about yourself deeply with the other members of your family depends on whether your parents talked personally to one another in front of you or talked personally to you. It seems that talking very personally isn't the usual thing for most families in the United States. In this country we tend to talk about more personal things to friends outside our families rather than to mother, father, brothers, or sisters. To whom do you disclose yourself most personally?

• *Fear of knowing yourself.* Self-disclosure, as Jourard suggested, is one of the principal ways we have of communicating not only with others but with ourselves. It's possible, then, that at times we're afraid to disclose ourselves to others because we don't want to get closer to ourselves. Self-disclosure can put us into contact with parts of ourselves that we'd rather ignore. For instance, in our culture good looks are given an exaggerated importance. It's very important to be attractive, and one of the reasons people fear getting older is that they become less physically attractive. If you're not very attractive, you might feel very bad about it and at the same time try to put it out of your mind. You don't admit your feelings to yourself or talk about them with anybody else. But they continue to hurt inside like a covered-up infection. If, on the other hand, you face the whole issue openly and say to yourself that you know that you aren't especially good looking but that this fact about you isn't really that important, then you're free to be yourself more. If you can also share your disappointment that you're not better looking with some people you trust and talk out your feelings, then you'll free yourself even more. You'll probably find out that looks are not as important as you've been making them. This group experience gives you the opportunity of getting to know yourself much better. Does this possibility make you a bit

47

nervous? If it does, you're probably not much different from your fellow group members.

• *Fear of closeness.* You can't reveal yourself on a deep level to another person without creating, by the very act of self-disclosure, some degree of closeness between you and that other person. Therefore, if you're somewhat fearful of self-disclosure, it may be that what you really fear is getting close to others. There are people who fear human closeness more than they fear death. Why should you fear closeness? Well, getting closer to others places certain demands and responsibilities on you. If you care about another person, you want to be available to him or her, and that puts limitations on your freedom. For some people, however, getting close to others just seems to be a scary thing. They don't know why, but they do know that their feelings keep them from getting too close. How do you feel about getting close to others? The laboratory gives you the opportunity to examine any fears you might have of closeness. The contract calls for forming and developing relationships in the group. We could add that it's better if these are relationships of some degree of closeness. The laboratory is safe because these relationships usually take place only within the group and only for a limited time. Some of the alternatives to human closeness are being isolated or lonely, having safe but superficial relationships with others, making work a substitute for close relationships, and playing interpersonal games with others. None of these alternatives is very satisfying.

• *Fear of change.* If you reveal yourself to another in any deep way, if you talk about the way you're living your life, including your interpersonal life, you may discover that you're not living up to the standards that you've set for yourself. You may discover that some of your values aren't really values at all but just good intentions or good ideas. These "values" that never get put into action are called *notional values*—they're good "notions" rather than real values. For instance, you may think that you're open to dealing or interacting with all sorts of people. However, as you talk with your fellow group participants and as you examine your actual interpersonal style with them (that is, as you discuss what you *actually* do instead of what you think you do), you might find that you have certain prejudices, that there are certain kinds of people you avoid dealing with. You might find, for example, that you make no attempt to contact people who hold opinions different from yours. You may find that it's difficult for you to form relationships with older people, whether men or women, or that you tend to ignore people you think are less intelligent than you are. If

48

you bring these discoveries up in the group, you almost automatically commit yourself to doing something about them; that is, you commit yourself to change. Either you have to change your values and say that being open to all different kinds of people is *not* a value for you, or you have to change your behavior. If you sense that you don't want to change in a certain area of your life or that you don't want to do the work required to change, it's quite likely that you won't reveal that area, that you'll keep it to yourself. It's much easier to forget about what you don't want to change than it is to change it. To the extent that this laboratory does demand self-disclosure, it also demands the possibility of change.

• *Fear of rejection.* Often people don't reveal themselves very deeply to others because they are afraid of being rejected. Many people think deep inside themselves "If others *really* knew me for the person I am, they wouldn't accept me." For instance, if I see that you're a person who works hard, while I know that I'm actually a rather lazy person, I may just keep my knowledge to myself because I don't want you to reject me. Or suppose that I'm annoyed because you tend to interrupt and bully me and others in the group. I don't tell you that this tendency annoys me, because I'm afraid that you'll get angry with me and reject me. Fear of rejection can make cowards of us all. Because of this fear, I may be afraid to share even my good points or values with you. For example, if cooperation is an important value for me, whereas you seem to value being independent and working on your own, I may say nothing to you about how I feel about cooperation. If you see that we're different in this respect, you may think less of me or reject me. My fear of rejection may also be related to my deeper feelings about myself: perhaps I fear that you won't accept me because, deep down, I don't like and accept myself.

One reason I may be afraid to tell you anything about my weaknesses is that I don't want to be seen as a weak person. If I tell you about a weakness in one area of my life—for instance, if I tell you that I'm afraid of people in positions of authority (teachers, policemen, my boss at work, my father)—you may think that I'm a weak person in all areas of life, which isn't true. In this case I don't trust that you will see me as a person who, like other persons, has both strengths and weaknesses, good points and bad. All of this points to the fact that a climate of trust has to be developed in the group if you and I are going to talk to each other very deeply. Later on in this book I'll discuss ways of developing trust in a group.

• *Fear of being ashamed.* All of us have experienced shame at one time or another in our lives, and most of us have found it an

49

uncomfortable experience. The word *shame* comes from a word meaning "to uncover, to expose, to wound." Although you may think of shame principally as something you experience when *others* find out that you've done something wrong, the experience of shame really starts at home. Shame isn't just being painfully exposed to someone else; it's first of all being painfully exposed to oneself. Shame is often a sudden experience: suddenly you feel deeply some inadequacy that you have, and shame flushes through your body and soul. You may be ashamed of something that you've done wrong, but you may also be ashamed of your looks or of the fact that you have some physical deformity or of the fact that you're not very intelligent. You can feel shame even though nobody else is around and even though nobody else realizes what you're feeling. But the feeling of shame is usually even more intense if you feel ashamed in front of others and they know that you're ashamed. Although it's true that others cannot make you feel ashamed unless you are first ashamed of yourself, still the presence of others makes the experience more intense and more painful.

Since shame is a kind of uncovering, it's related to self-disclosure, which is also a kind of uncovering. Therefore, you may fear self-disclosure because it might lead to an experience of shame. As we've already seen, self-knowledge can be painful, and self-disclosure often leads to deeper self-knowledge, which in turn can lead to shame. But shame need not be just a negative experience. In fact, it can be a much richer experience than you might think. Feelings of shame need to be recognized and faced, not avoided. It's possible that an experience of shame, if faced, will throw unexpected light on who you are and point toward the better person you might become.

A young man came to me once to talk about an experience of shame. He'd been playing basketball. In a game of "shirts" and "skins," for the first time in his life he had been a "skin." When he took off his shirt, someone remarked that he wasn't very well built. Suddenly he had an intense experience of shame. He was ashamed of his body; he was ashamed of himself. Nobody else on the basketball court noticed, but that didn't lessen his experience of shame (remember that shame is primarily being painfully exposed *to oneself*). He also found it difficult to talk to me about this experience. But, as he explored his feelings, he came to realize that without knowing it he'd been buying certain values in society that he really didn't want to. He'd been saying to himself that good looks, including a good physique, is a high value. But, just as he didn't

50

want to accept or reject others just because of their looks, so too he didn't want to reject *himself* because of his looks or because of the kind of body he had. In his case, facing the experience of shame did throw unexpected light on the person he was. It showed him that he was holding values he didn't want to hold, and it pointed toward the kind of person he wanted to become.

You may or may not experience some shame during this laboratory. The purpose of the laboratory and of self-disclosure is not to make you feel shame. If there is some experience of shame, however, it can be growthful, especially if it's talked about with your group as members of a learning community—that is, with a group of people you can trust.

Appropriate Self-Disclosure

As we've already seen, the word *appropriate* means *fitting* or *suitable*. If something is inappropriate, it's out of place. Therefore, appropriate self-disclosure means the right kind and the right amount at the right time. In order to judge whether self-disclosure is appropriate or not, you should take into consideration such things as:

- *amount.* How much information is being disclosed?
- *depth.* How deep or intimate or personal is the information?
- *time.* How much time is spent disclosing yourself?
- *the person talked to.* To whom are you revealing this information?
- *the situation.* Under what conditions or in what situation are you revealing the information?

Consider a young man who is constantly (time) talking about himself in a very superficial way (depth) to anyone who will listen to him (the person talked to) on almost any occasion (situation). Such a person is usually very boring. Although he says many things (amount), he really says nothing. Chances are that people will usually try to avoid him. His self-disclosure is inappropriate.

Self-disclosure and mental health. The factors listed above (amount, depth, time, the person talked to, and the situation) have been studied, and an attempt has been made to relate them to the mental health or adjustment of the person who does the disclosing. The evidence shows that there are differences between well-adjusted people and poorly adjusted people in the area of self-disclosure. Well-adjusted people tend to disclose a great deal (amount) of very personal things (depth) to a few close people (the person talked to). They disclose a moderate amount of moderately personal things to other people, such as their ordinary friends and acquaintances; that is, they disclose enough to keep the relationships going but not enough to frighten these people. On the other hand, poorly adjusted people tend to be either overdisclosers or underdisclosers to practically everyone. Poorly adjusted people have a hard time figuring out how much to say and when to say it. They talk about very personal things in public or with complete strangers, or they say practically nothing about themselves to anyone.

Setting up standards for self-disclosure. The standards we set up to determine whether something is right or wrong, good or bad, appropriate or inappropriate are called *criteria.* We use criteria to judge things, to put them in one category or another. Criteria or standards will now be set up so that you can judge the appropriateness of your self-disclosure. These standards will help you in disclosing yourself both in the group and in everyday life. I'll give examples in order to make the standards or criteria as clear as possible. It isn't necessary to memorize either the standards or the examples. However, they can serve as a kind of checklist for your disclosure in the group. If, for instance, someone says to you "You really haven't said very much about yourself in this group," you can use these standards to check whether what that person is saying is right or not.

1. *Self-disclosure should be directed toward reasonable goals.* Your disclosures will probably be more appropriate if you know why you're disclosing yourself, what you're trying to achieve. In the laboratory group you're disclosing things about your interpersonal style in order to examine that style. When people talk about themselves just for the sake of talking about themselves, they generally become either exhibitionists (people who reveal too much at the wrong time, either physically or psychologically) or bores. In the laboratory you're trying to form and develop relationships with your fellow group members. Therefore, self-disclosure that helps you do precisely that is most useful. Consider the following example.

> "I'm a shy person. I find it hard talking to strangers and talking in groups. My words get all mixed up, and I forget what I was going to say. At times I get so scared that I can't even join a conversation. Well, here I am—in a group of

strangers—and I'm finding it difficult to get going. I think it's only fair to let you know what's going on inside me."

Since the goal of the group is to form and develop relationships in order to examine interpersonal styles, this person's self-disclosure is quite appropriate. He tells the other members of the group what he's like in certain interpersonal situations and at the same time reveals something that is getting in the way of establishing relationships with them. His self-disclosure helps the other members to give him the support needed to speak out in the group.

2. *The depth and the amount of your self-disclosure should be related to your goals.* Your self-disclosure will have a greater impact if the *depth* and the *amount* of the disclosure are also related to your goals. If your goal is to establish a close friendship with someone, then your self-disclosure will have its greatest impact if it gradually becomes more and more personal and extensive—that is, if you reveal not only more things about yourself but also more intimate things. In the laboratory group, however, most of your fellow group members are probably not going to become your very close friends, even though your immediate goal is to establish and develop relationships with them that have some degree of closeness. Therefore, it's up to you to determine, in the flow of the group conversation, how personal you want to be in your disclosures. The goal of the group is not dramatic self-disclosure, or "secret-dropping." The group isn't a place where people play a game in which whoever drags the most and most gruesome skeletons out of the closet wins. This kind of disclosure is simply not related to the goals of the group. On the other hand, merely shallow or superficial disclosures about yourself aren't related to the goals of the group either. If you really want to establish relationships in order to examine your interpersonal style, you'll have to take some *reasonable* risks in self-disclosure. It's impossible to set down exact rules concerning how much you should reveal and how personal your disclosures should be. The standards or criteria we're now discussing and the examples presented throughout this book will give you some help, but learning how to direct your disclosures to the goals of the group is also something you'll learn by doing.

3. *Respect and caring should be shown in the giving and receiving of self-disclosure.* It's difficult, if not impossible, to disclose yourself to people you don't care about or respect. If you respect the members of your group and have a basic desire to relate to them, self-disclosure will be a means of improving your relationships with them.

Self-disclosure takes place most easily in a climate of trust and support. If you expect others to disclose themselves to you, it's necessary for you to let them talk about themselves without fear of being judged. On the other hand, if you want to disclose yourself to them, you'll probably do so only if you expect that they will give you a fair hearing. But you'll also disclose yourself only if you think that it's worthwhile to talk about yourself to this particular person or group of people. Most of us have certain unexamined interpersonal prejudices, ones we haven't admitted even to ourselves. Your emotions might tell you that you really don't like to relate to older (or younger) people, to people who seem too straight (or too far out), to religious people (or people who aren't religious at all), to blue-collar workers (or white-collar workers), to people who are too emotional (or too cold). If these little prejudices get the best of you from the beginning, you'll find it very hard to disclose yourself. Your self-disclosure in the group will gradually become deeper only to the degree that your respect for the others becomes deeper too. Otherwise, either your disclosure won't be appropriate or you'll simply refuse to disclose yourself.

4. *Continuing relationships call for self-disclosure.* Self-disclosure is appropriate if the relationship you have with another person has some kind of future. There *should* be a difference between the kind of disclosure that takes place in the group and the kind of disclosure that takes place between you and your best friend. On the other hand, your disclosure in the group shouldn't be superficial just because these people aren't going to be your lifetime friends. In fact, if you and your fellow group members establish a climate of caring, cooperation, support, and trust, then your disclosures might be even *deeper* than in your everyday life, where the kind of support you receive in the group doesn't exist. The serious work you and your fellow group members are doing in the group demands serious relationships. And serious relationships call for serious self-disclosure.

5. *There should be give-and-take in self-disclosure.* Evidence from psychological studies shows that, if you talk personally about yourself to other people, they will tend to talk personally about themselves to you. Maybe you don't need a psychological study to tell you that, because you see it happening in your interpersonal life. Self-disclosure is inappropriate if it's one-sided, if there's no give-and-take. What will happen to our relationship if I reveal myself to you but you don't say anything about yourself to me? The relationship will probably begin to fall apart. If I tell you about myself, and if you merely listen and make comments on what I say, then you become my counselor or my helper. The laboratory group, however, isn't a counseling

group. You're not in the group to confess your sins to a priest or to tell your problems to a counselor; in fact, that kind of disclosure would be inappropriate in the group. Therefore, if you disclose yourself, you have the right to expect that the others will do the same. The person who doesn't disclose himself or herself in this kind of laboratory group usually ends up being ignored.

6. *The timing of self-disclosure is important.* Self-disclosure is more natural if it grows out of the give-and-take of the group's conversations. Since self-disclosure isn't a goal in itself but rather a means of establishing relationships through which you can examine your interpersonal style, dumping what you have to say about yourself like a truckload of sand is inappropriate. Relationships are built up gradually, and this process takes a certain amount of patience. In the laboratory group we do try to speed up the process of establishing relationships, and both the contract and certain group exercises that will be presented later on are some of the means we use in this effort. Nevertheless, "speeding up" works only to a certain degree; it doesn't eliminate the need to establish and deepen trust among the group members, and this process does take time.

Timing is also important in another sense. For instance, disclosing how you feel, if you do it at the right time, can help move the group forward. Let's take an example of a time in the group when nobody seems to know what to say. During this lull a young woman says:

> "We seem to have a hard time getting in touch with one another. At least I am. And I think I know why I feel awkward and don't know what to say. I'm still not too sure how personal and deep I want to be in here. Not that I have anything against you people. I'm the same outside. Getting close to people takes a lot of work, and I'm just beginning to realize how lazy I am. I don't usually work at making closer friends. I do more work at relating in here than I do outside."

This member uses a time when the group itself seems to be drifting to talk about her own hesitancy or reluctance to work harder. What she says hits the mark because it's related to what is happening (or what is *not* happening) in the group. The timing is good.

7. *Self-disclosure should be related to the here and now.* If you and I are in a group together and I begin talking about what happened to me at work last month, I'm not talking about the here and now but about the there and then. The here and now refers to what's

happening in *this* group at *this* time. In human-relations-training groups, self-disclosure is usually appropriate to the degree that it's related to the here and now. This doesn't mean that you may not talk about your life outside the group; such a restriction would certainly make the group very artificial. However, it does mean that, when you talk about matters from the past (the "then") or matters outside the group (the "there"), you should make an effort to relate the there and then to the here and now—that is, to relate past or present experiences outside the group to what is taking place *now* in *this* group. Here are some examples of how there-and-then material can either be a distraction, a way of avoiding the work of the group, or a significant contribution to what is happening in the group in the here and now.

There-and-Then Material

Inappropriate because it remains there and then

"My dad and I don't get along. He's usually down on my brother, too. It makes living at home very difficult. Sometimes I get so angry that I want to just move out."

"Last year I didn't do well in school or in my social life. I failed a course. I was lonely. It seems I couldn't get it together."

Appropriate because it is related to the here and now

"Tom, I'm reacting to you just the way I do to my dad. He doesn't listen to me, and I'm not sure that you do either. I feel like turning my back on you in here, just as I feel like leaving my dad and getting an apartment of my own."

"Last year was a miserable year away at school. I was lonely and did poorly at school. I was afraid of so many things . . . and I still have the same fears. I'm afraid I won't do well here, so I don't say anything, and my fears just get worse. I'm just beginning to see that most of my fears are stupid. I *let* them get the best of me."

In each of these cases the person talks about what is currently taking place outside the group or what has taken place in the past. These disclosures become appropriate to the degree that they are related to the here and now of the group—to the process of establishing and developing relationships and to the goal of examining interpersonal style. Dwelling on the there and then has a way of turning a human-relations-training group into a counseling group.

8. *Reasonable risks with self-disclosure should be taken.* Taking risks and creating trust in the group are related to each other. It seems like a problem impossible to solve: in order to take risks it's necessary to have a climate of trust, but in order to create a climate of trust it's necessary to take risks. In practice, reasonable and gradual risk taking in self-disclosure does a great deal to develop an atmosphere of trust. A group in which all disclosures are very safe soon becomes boring. On the other hand, a person who reveals too much too quickly frightens others, and they tend to fall silent. During the first meeting of a group I was in, one of the members revealed, with great pain, how he'd been fired from teaching in a high school some years back. He had been involved in some kind of scandal. He talked about his pain and his embarrassment and relived some of the pain and the embarrassment in the group. This disclosure was too much too soon for the other members of the group. They listened to him and made a few remarks, but during the next meeting they attacked him for dumping too much on them from his past. They were angry because he'd risked too much too soon. They accused him of trying to set a standard of self-disclosure that was too high. During the rest of the group meetings they kept referring to his "bomb." In this case, at any rate, the risk was seen as too big too soon, and it made the other members of the group overly cautious in self-disclosure. They used his "bomb" as an excuse for not revealing very much about themselves.

A Checklist on the Appropriateness of Self-Disclosure

Here is a checklist that can help you prepare your self-disclosure as you're writing your log and agenda for each meeting:

- If I'm a bit slow in disclosing myself in the group, is it possible that I'm fearful of getting to know myself better?
- If self-disclosure is difficult for me, is it because getting closer to others is difficult for me?
- Do I hold back from disclosing myself because I'm fearful of being ashamed?

- Am I hesitant in talking about my behavior because I have mixed feelings about changing some of my behavior?
- Is self-disclosure hard for me because I think that people might reject me if they know me well?
- Is my self-disclosure aimed at establishing relationships and examining my interpersonal style?
- Am I talking personally enough about myself, or am I staying too close to the surface?
- Do I talk too much about myself, not giving others a chance and not listening and responding to them enough?
- If I'm a bit slow in talking about myself, is it because I don't care about or respect the members of my group enough?
- Am I slow in self-disclosure because I'm afraid that the others do not respect or care about me?
- Are the relationships with the others in the group important enough for me to do the work necessary for good self-disclosure? Do I want these relationships even though they're temporary?
- Is there good give-and-take in the group? Do we have only a couple of people who are good self-disclosers, or is everybody trying?
- Do I prepare my self-disclosure so that I not only know, in general, what I want to say about myself but also know that my self-disclosure is timed right, that it fits?
- Am I moving forward in my self-disclosure, neither going too quickly nor dragging my feet?
- Is my self-disclosure related to the here and now, or do I talk too much about the past or about what takes place outside the group without relating it to my fellow group members?
- Am I taking reasonable risks in self-disclosure? During the group meetings do I keep a lot of thoughts and feelings to myself, or do I share them willingly with the others?

Answering these questions for yourself is a way of deepening your self-awareness. As we've seen, deepening your self-awareness is the first step in developing the skill of self-disclosure. In different ways you ask yourself "What's going on inside me that keeps me from making deeper contact with other people by revealing myself to them?"

Responding to self-disclosure. As you will see in Chapters 6, 7, and 8, self-disclosure is much easier in a group in which the members listen carefully to one another and show one another that they understand. The best immediate response to someone else's self-disclosure is not self-disclosure on your part but rather understanding. Responding with understanding enables the self-disclosing person to say "Oh, someone has listened to me and understands what I'm trying to say. That's a good feeling." If one of the group members' self-disclosure is met by silence, or if some other member changes the subject without trying to understand, then the self-disclosing member might think twice before trying again.

A balanced picture of yourself. As I've said a number of times, self-disclosure isn't an end in itself. When some people think about self-disclosure, they immediately think about their weaknesses or about things they think they've done wrong. Self-disclosure that consists almost entirely of some kind of confession of weakness is inappropriate in the group. If you are to get a deeper understanding of your interpersonal style, you have to come to know your strengths and good points just as well as your weaknesses. So the laboratory is a time for exploring and developing your strengths and not just a time for examining and confessing your weaknesses. *Groups that spend too much time exploring weaknesses instead of developing strengths become too heavy and depressing.*

The Different Ways in Which You Disclose Yourself to Others

I've been talking about self-disclosure in a way that might seem to imply that you can *decide* whether you want to disclose things about yourself or not. The truth is that you're *constantly* disclosing yourself, because your self-disclosure isn't limited to what you *say.*

Verbal self-disclosure. Of course you can use words to let others know about you. You can state in words what people already know because they spend time with you ("I'm a rather shy person"), or you can tell them through words what is hidden from them ("I daydream a lot about being a successful musician and being liked by a lot of people"). But there is more to self-disclosure than just what you say about yourself.

59

Body-language and tone-of-voice self-disclosure. People may know things about you, even things you would prefer to hide, even though you never say anything about these things in words. Your body, your gestures, the tone of your voice, the way you look at another person all say something about you. For instance, if someone invites you to a movie and you say yes, but slowly and with little enthusiasm, you really say that you don't want to go. The real message is not in the word "yes" but in your tone of voice. Part of your interpersonal style is your verbal behavior—how you use words in talking with others, what you have to say. But a great part of your interpersonal-communication style is your nonverbal behavior and your tone of voice—the messages you give off that color what you say with your words. It's important that you get feedback in this laboratory on your entire interpersonal style, including how you communicate without using words.

Self-disclosure through actions. What you *do* tells others a great deal about you. Some of the actions that say a lot to others about you are:

- how you dress,
- the kinds of friends you choose,
- the kind of work you do,
- what you do with your free time,
- what you read, and
- what you watch on television.

What other actions would you add to this list? Actually, all of your actions say something about you; the lab gives you a chance to ask yourself *what* they say about you.

Since you're disclosing yourself all the time—if not in words, then through your actions and your body language and tone of voice—you're already a very public person. Certainly you have private information about yourself that others will have to know if they're going to understand the full meaning of your public behavior. For instance, suppose that, when you talk to others, you say very little. Your words are short and your manner abrupt. You spend a great deal of time by yourself. When others "read" your behavior, they think that you don't care for them or perhaps for people in general. Inside, however, *you* know that you're a lonely person. You're short with people and you stay away from them mainly because you're scared. In this case public information—your words, your

60

body language or nonverbal behavior, your actions—isn't enough to give a true picture of you.

Exercises: Self-Awareness in Preparation for Self-Disclosure

The purpose of most of these exercises is to help you think about yourself and especially about your interpersonal style. After doing an exercise, there should be a lot that you *could* say about yourself. However, what you *do* say about yourself is still up to you. Exercises help you prepare yourself for the group discussions. They help you prepare your agenda for the group. But they shouldn't become a substitute for spontaneity in the group. Do you remember the example of the basketball player who practices hard? By the time the game comes around, he or she has many different moves and thus has the ability to be spontaneous on the court—to use whatever move is best at any given time. These exercises will help you develop many different moves in terms of self-disclosure, but the way you move at any given time in the group will depend on what's happening in the group. Your instructor will also give you suggestions concerning ways to use the information you discover about yourself through these exercises.

Exercise 5
The "I Am" Exercise

In this exercise you're asked to take the sentence "I am . . ." and finish it in 20 different ways. Try to complete each sentence in ways that will help you to get at your interpersonal style or at qualities that affect your interpersonal style.

Example

- I am . . . intelligent.
- I am . . . a bit scared when I meet new people.
- I am . . . very changeable, liking people for a while and then dropping them.
- I am . . . fun loving.
- I am . . . seductive in various ways.

61

- I am . . . usually enjoyable to be with.
- I am . . . very understanding, when I want to be.
- I am . . . often bored with everyone and everything.
- I am . . . not very caring at times, because I put my comfort ahead of people.
- I am . . . not very pleased with myself at times.
- I am . . . good looking enough.
- I am . . . too loud-mouthed with my friends.
- I am . . . a puzzle to myself at times.
- I am . . . very open to new people once I get to know them a little.
- I am . . . quite open with older people and with kids.
- I am . . . not sure about sex in my life.
- I am . . . interested in the Peace Corps.
- I am . . . stingy with money, though I don't like others to know it.
- I am . . . willing to change the ways in which I relate to others, especially my friends.
- I am . . . interested in religion, but I'm afraid to talk to anyone about it.

Once you've written out your own list, look at the checklist on appropriateness of self-disclosure (pp. 57–58) and see which of the items are appropriate to talk about in your group.

Exercise 6
Self-Disclosure Cards

As you do the "I am" exercise, you probably think about things concerning yourself and your interpersonal style that you feel will fit into the group discussions and that you're willing to talk about. It's very helpful to keep the general ideas of what you want to say about yourself on small (3-by-5-inch) file cards. When you begin practicing interpersonal skills with your fellow group members, you'll be asked to talk about yourself. You'll find the task much easier if you're prepared. You'll be prepared if you write down ideas about what you want to say about yourself whenever you do these exercises—or, for that matter, whenever you think of something you'd like to say. You can use the cards to prepare your agenda for the next meeting (part of your agenda should always be what you'd like to say about yourself), and you can bring the cards with you to the skills-practice sessions and use them to prompt yourself when you're asked to talk about yourself.

Developing themes. A self-disclosure theme is a general topic that you can develop more fully and more concretely as you talk about yourself. As an example, one theme might be being cautious or careful in interpersonal situations.

- Theme: "I'm very cautious about the ways I relate to others."
- Development of the theme:
  - "When someone gives me signs that he or she likes me, I don't react for a while. I think that some people take this to mean that I don't like them, and they give up."
  - "I choose my words very carefully when I'm talking to someone I like. At times they must think I'm too silent or slow."
  - "I hesitate to call people up at times because I'm afraid that I'll bother them. I hate to ask someone to do something and get turned down."

Here, the general theme or topic is being careful (or too careful) in relating to others. Developing the theme means seeing how the theme affects your interpersonal behavior. You'll find it useful to come up with some themes that are important to you. The group is a good place to examine these themes, especially if they affect your relationships within the group itself. Important themes can be written down on file cards.

Exercise 7
Incomplete Sentences

This is another exercise to help you think about your interpersonal style and develop interpersonal themes.

Directions

The exercise is a simple one. Just finish each incomplete sentence. Do the exercise relatively quickly; that is, don't spend a lot of time thinking what you will or should put down. Write down whatever comes naturally to you. There are no right or wrong answers.

63

1. People who love me ...
2. One thing I really like about myself is ...
3. I dislike people who ...
4. When people ignore me ...
5. The way I'm generous with others is ...
6. When someone praises me ...
7. When I relate to people, I ...
8. When I relate to people, they ...
9. Those who really know me ...
10. When I let someone know something I don't like about myself ...
11. My mother ...
12. My moods when I'm with others ...
13. I'm at my best with people when ...
14. When I'm in a group of strangers ...
15. I feel lonely when ...
16. I envy ...
17. When someone is affectionate with me ...
18. When I take a good look at my interpersonal life, I ...
19. The way I handle jealousy is ...
20. I think I've hurt others by ...
21. Those who don't know me well ...
22. My brother ...
23. The person who knows me best ...
24. An important interpersonal value for me is ...
25. What I'm really looking for in my relationships is ...
26. I get hurt when ...
27. I daydream about ...
28. My family ...
29. When someone confronts me ...
30. I'm at my worst with people when ...
31. What I feel most guilty about in my relationships with others is ...
32. I like people who ...
33. When someone gets angry with me ...
34. My sister ...
35. Few people know that I ...
36. When I think about closeness, I think about ...
37. When I meet someone who is very strong and outgoing ...
38. When I don't like someone who likes me, I ...

39. When I'm not around, my friends . . .
40. Most people think that I . . .
41. One thing I really dislike about myself is . . .
42. When I'm with a group of my friends . . .
43. I get angry when someone . . .
44. What I distrust most in others is . . .
45. One thing that makes me nervous in interpersonal situations is . . .
46. When I really feel good about myself, I . . .
47. When others put me down . . .
48. In relating to others, I get a big lift when . . .
49. In my interpersonal relationships this year I learned that . . .
50. When I like someone who doesn't like me, I . . .
51. I feel awkward and out of place with others when . . .
52. When others act like my parents towards me, I . . .
53. The thing that holds me back in my relationships with others is . . .
54. Too many people . . .
55. When I share my values with someone . . .
56. I would like the person I marry . . .
57. Others like it when I . . .
58. Interpersonal relationships are important, but . . .
59. When others see the ways in which I can be hurt . . .
60. In interpersonal situations what I run away from most is . . .

Once you've completed the sentences, go back and circle the ones that you think are most important, the ones that say a lot about your interpersonal style. Put an *X* in front of statements that you *don't* want to disclose to the members of your group. Use the remaining circled statements to help you fill out your self-disclosure file cards.

You may want to return to this exercise later on. Statements that you think are important but that seem to be inappropriate self-disclosure right now can be kept on file. As you get to know the members of your group better and begin to trust them, you may be willing to reveal deeper things about yourself. In general, use the checklist on appropriateness of self-disclosure to determine what you want to say about yourself in the group.

65

Exercise 8
Mapping Your Interpersonal Style

Here is another exercise to help you think about and explore your interpersonal style. This exercise can help you become more concrete and specific about what you want to say about yourself.

Directions

Read each pair of statements, and decide which one describes you or your interpersonal style better. Next examine the rating sheet at the end of this exercise. For each pair of statements, once you decide whether you choose *a* or *b*, then decide just *how well* the statement you choose fits you—that is, how well it describes your interpersonal style. For instance, if in the first pair you choose statement *b* as describing you better ("I'm an outgoing person. I enjoy meeting new people. I even look for opportunities to get to know people."), then you should put an *X* in the column that for you indicates the best "fit." Therefore, if you think that this statement fits you perfectly, mark an *X* under "very much like *b*."

Statements on Interpersonal Style

1. a. "I'm shy. My shyness takes the form of being afraid to meet new people and of being afraid to talk about myself deeply with the friends I do have."
   b. "I'm an outgoing person. I enjoy meeting new people. I even look for opportunities to get to know people."
2. a. "I don't push myself enough in interpersonal situations. Others can step on me and I usually take it without saying much."
   b. "I stand up for my rights fairly well. I'm kind to others, but I don't let them step on me or control me."
3. a. "I get angry very easily, and I dump my anger on others freely. I often get angry because I don't get my own way."
   b. "Although I become angry at times, I don't lose control. When I'm angry with someone, I tell the person so and try to settle what's bothering me."

4. a. "I'm a lazy person. I find it especially hard to take the time and energy needed to get to know others."

   b. "I'm a very energetic person. I like listening to and getting involved with others. I work hard at the relationships I have."

5. a. "I'm somewhat fearful of persons of the opposite sex, especially if I get the feeling that they want some special kind of closeness with me."

   b. "I get along very well with persons of the opposite sex, and I can have many different kinds of relationships with them, from casual friends to someone very close."

6. a. "I'm not a very sensitive person. I find it hard to know what others are feeling, and often I don't care."

   b. "I'm usually aware of what others are feeling. Often I find myself experiencing something of the same emotion as someone else, just by listening to them."

7. a. "I have too much self-control. I ordinarily don't let my emotions show at all. Sometimes I think that I would rather not have emotions."

   b. "I express my emotions well. I don't dump them on others, but I don't try to hide them either. Others usually know what I'm feeling."

8. a. "I like to control others, but I don't want others to know that I'm doing it. In my relationships I want to be the person in charge."

   b. "There is a great deal of give-and-take in my relationships with others. I don't let others boss me around, but I don't have to be the person in charge either."

9. a. "I have a very strong need to be liked by others. I want everyone I meet to like me. I seldom do anything that would offend anyone."

   b. "I like to be liked by others, but that isn't the most important thing in my life. I want to do what I think is right, even if others don't like it. I don't like to offend others, but I don't need everyone's approval either."

10. a. "I feel that I just have to help others. I get nervous when I'm not doing something for someone else."

    b. "Helping others is important, but it's also important for me to have time for myself. I don't need people

67

who need my help or who want me to do favors for them in order to feel good about myself."

11. a. "I'm very easily hurt. I send out messages to others that say: 'Be careful of me.'"

  b. "I don't keep thinking that others might hurt me. I roll with the punches pretty well. I can laugh at myself. Others know that they can be loose when they're around me."

12. a. "I don't want to be dependent, so I fight others. I always have to show that I'm free and a person in my own right. I find it quite difficult to get along with people in authority."

  b. "I like cooperating and working with others. I can influence others and let others influence me. I can depend on others and let them depend on me in a good sense."

13. a. "I'm an anxious person, especially in interpersonal situations. I don't know why I'm like that."

  b. "I'm a relaxed person. Although I do get anxious at times, it's more like me to be relaxed in interpersonal situations."

14. a. "I'm a somewhat colorless and uninteresting person. I have few interests. I'm bored with myself at times, and I assume that others are bored with me."

  b. "I'm a colorful and interesting person. Others enjoy being with me. I add life and excitement at gatherings, and I like this part of myself very much."

15. a. "I take too many risks, and I'm too daring in interpersonal situations. I'm impulsive. I lack self-control."

  b. "I take reasonable risks in interpersonal situations. I have fairly good self-control."

16. a. "I'm very stubborn. I think my opinions are more important than anyone else's. I'm ready to argue with anyone at any time."

  b. "I have an open mind. While I have ideas of my own, I don't go around looking for arguments. I enjoy sharing my opinions with others and can even change my opinions when I see something better."

17. a. "I'm a rather sneaky person. I seduce people in various ways (not necessarily sexual ones). I get them to do what I want."

   b. "I'm open and direct in my relationships with others. If I want anything from anyone, I ask them for it as plainly as possible. I'm not sneaky or seductive."

18. a. "I'm a selfish person. I usually put my own needs and comfort before the needs of others."

   b. "I can be generous when I want. There are times when I do put the needs of others before my needs and comfort."

19. a. "I feel awkward when I'm with people. I don't do the right thing at the right time. I don't notice when others are having a rough time."

   b. "I'm very much at home with people. I'm sensitive to what those around me are feeling, and I usually respond to what they're feeling in a human way. I usually know the right thing to do at the right time."

20. a. "I'm rather lonely. I don't think that others really like me. I spend a lot of time feeling sorry for myself."

   b. "I experience loneliness from time to time, but it's not a big thing. I know that people think that I'm all right. I don't spend much time at all feeling sorry for myself."

21. a. "I'm stingy. I don't share my money or my time very easily."

   b. "I'm a generous person. I like to share what I have with others, and I like to receive what others have to share with me."

22. a. "I've lived too sheltered a life. I feel out of it; I don't really know the score on life. Others have had many more experiences than I've had. They know life a lot better."

   b. "I've had plenty of human experiences. I know the score just as well as the next person. I'm not just a kid."

23. a. "I'm something of a coward. I find it hard to stand up for my opinions or convictions. It's easy to get me to retreat."

   b. "I have the courage of my convictions. It's not easy to bully me. I know what my values are, and I can stand fast with them even when others challenge me."

24. a. "When I'm challenged or confronted, I tend to run away or else to attack the person confronting me. I get scared."

b. "When I'm confronted, I listen to what the other person has to say. I try to make sure I'm understanding what he or she says. I try to respond without being defensive."

After you've filled out the rating sheet, turn it on its side so that the *b*s are on top. As you've probably noticed, the *a* statements are negative, the *b* statements positive. If you draw lines connecting each *X*, your rating sheet becomes a kind of graph, the hills indicating your strong points and the valleys your weaknesses. This exercise is meant to help you get a *balanced* picture of yourself. In the group experience you can find out whether your judgments about yourself are accurate, both by watching yourself and by getting feedback from your fellow group members. An accurate view of your strengths and weaknesses is necessary if you want to try to change some parts of your interpersonal style for the better. Some of the questions you might ask yourself as you study your graph are:

- Where are the hills, and where are the valleys?
- Do I find any self-disclosure themes or topics here?
- Are there a couple of *X*s that stand out, whether positive or negative?
- Am I too hard on myself? Too easy?
- How can I develop my strengths?
- What can I do about my weaknesses?
- Taking the rules of appropriate self-disclosure into account, what have I found out about myself that I'd be willing to share with my fellow group members?

Anything you discover through this exercise that is worth talking about in the group can be put on your self-disclosure file cards.

70

RATING SHEET

I am: very much like *a*.	fairly much like *a*.	a little bit like *a*.	*I am:* a little bit like *b*.	fairly much like *b*.	very much like *b*.

1a.						1b.
2a.						2b.
3a.						3b.
4a.						4b.
5a.						5b.
6a.						6b.
7a.						7b.
8a.						8b.
9a.						9b.
10a.						10b.
11a.						11b.
12a.						12b.
13a.						13b.
14a.						14b.
15a.						15b.
16a.						16b.
17a.						17b.
18a.						18a.
19a.						19b.
20a.						20b.
21a.						21b.
22a.						22b.
23a.						23b.
24a.						24b.

Expressing Feelings and Emotions

4

This chapter discusses the part that feelings and emotions play in your communication with others. Feelings and emotions can disrupt your interpersonal life, or they can enrich it—if you take responsibility for your own emotions and make them work for you instead of against you. Emotional control, assertiveness, and risk taking are among the topics discussed.

Some of us have trouble expressing our feelings and emotions. Either we keep them too much in check or, when we do express them, we express them poorly. As a result, our emotions stand in the way of better relationships with others instead of helping these relationships. For instance, you may never get angry or at least may never let others see your anger. Perhaps you don't think it's right to show anger, or perhaps you're afraid to show it because things will get out of control. On the other hand, you may express lots of feelings and emotions, but in ways that put people off. For instance, you may be the type to blow up at others when you get angry, with the result that they begin to avoid you—at least when your temperature seems to be rising.

It's unfortunate that we're never really taught how to show emotion in ways that *help* our relationships. Instead, we're usually told what we should *not* do. However, too little emotion can make our lives seem empty and boring, while too much emotion, poorly expressed, fills our interpersonal lives with conflict and grief. Within reason, some kind of *balance* in the expression of emotion seems to be called for. I'm not suggesting that we can program our feelings and emotions so that they can be turned on or off at will, but I *am* saying that there are ways of making feelings and emotions a colorful part of our lives. If we're not, within reason, the masters of our emotions, they'll probably master us. When we push all emotion out of our lives, we admit by that very act that we can't handle them.

Some Standards for Expressing Feelings and Emotions

The expression of emotions differs from self-disclosure. What you disclose about yourself is usually under your control. In the group it may happen that others are revealing themselves deeply and that you feel almost called upon to engage in some deeper self-disclosure yourself. Even in this case, however, you have the right not to. Emotions, on the other hand, aren't so easily controlled. If someone does something to make you angry, it's a natural response to get angry and show it (unless you've learned to keep all of your emotions under wraps). There's something almost automatic about emotions. And yet we've all learned how to control our expression of emotions on certain occasions. If someone makes a grammatical mistake while delivering a talk in front of the class, you might find it very funny and want to laugh but still control yourself because you don't want your fellow student to be embarrassed. Although you feel the emotion inside, you're able to control the expression of it.

In the last chapter I talked about certain standards or criteria for appropriate self-disclosure. In this chapter I'm going to discuss certain standards or rules or criteria for expressing emotion. During this discussion it would be helpful for you to remember that emotions aren't under our control in the same way that self-disclosure is. These standards, then, aren't rigid rules; rather, they're *guidelines*, suggestions for making feelings and emotions a richer part of our interpersonal lives.

1. *Emotions are OK.* Experiencing feelings and emotions is an important part of being human. No one has to apologize for experiencing them. The fact that they can be dangerous or misused doesn't mean that they aren't all right in the first place.

Yet feelings and emotions will always seem inappropriate to you if you can't accept them as a part of being human. The way you look at emotions, especially as they occur in interpersonal situations, is something you've *learned* in growing up. Both the way you look at emotions and the ways in which you express them or refuse to express them have become part of your interpersonal style. You can use some of your time in this group to see what you've learned about emotions. Do you enjoy deep emotion in interpersonal relationships? Do you try

74

to avoid emotional situations? Are there some emotions you fear, such as anger, strong sexual feelings, or depression? Part of the learning that you may have to do about emotions is discovering that they are a legitimate part of human living. Once you accept emotions as part of being human, you can decide just what role you want them to play in your life.

2. *Genuine emotions.* The purpose of this laboratory group is not to get you to manufacture emotions. Made-up emotions aren't genuine. If emotions *do* come up as you interact with the others in your group, however, you *are* asked not to hide them, even if hiding them is what you might ordinarily do. If you feel hurt by something that happens to you in the group, see whether you can let others know that you've been hurt. To do so may not be part of your present interpersonal style, but letting others in on your emotional life, especially insofar as it affects your participation in the group, is one way to experiment with new behavior. Sometimes some group members will want you to feel and express the same emotions that they are feeling and expressing. But you have your own feelings and emotions, and they're legitimate even if they aren't the same as those of others.

3. *A variety of emotions.* Some people become specialists in certain emotions. They're very good, perhaps too good, at expressing anger, say, but not so good at expressing joy. Many people have more trouble expressing positive emotions. They can cry, express grief, or feel sorry for others; they can express how they feel when they're down in the dumps, hurt, or rejected; but they can't express joy, affection, enthusiasm, peacefulness, satisfaction, and the like. Other people are just the opposite—they can't express negative emotions. Do you have any trouble with either set of emotions? Part of your interpersonal style may be to specialize in certain kinds of emotion. People may say of you "You can certainly tell when Cathy's in the dumps, but I'm never sure when she feels good about something." The lab gives you an opportunity to give expression to many different kinds of emotion. Try to see whether you can use this opportunity to experiment with the emotions you generally avoid or hold in. At first you may feel awkward expressing them, but by doing so you can decide whether you want to make them a part of your interpersonal style.

4. *Using feelings and emotions to promote the goals of the group.* Emotions can be used as weapons or as ways of controlling others. For instance, you might say to yourself "I don't like what Jim just did. I'm going to let him have it." In this example, anger is just a club and probably won't do much to help the group reach its goals. Or you

might say to yourself "If I cry, the other members will feel guilty and leave me alone." Here you're using emotion to control others. Both positive and negative emotion can be put to much better use. For instance, you might say in the group:

> "I feel pretty cautious and scared right now. Two people have just revealed a lot of personal things about themselves, and I'm not sure anyone really listened to them. At least no one has said anything to them. I think they're still out on a limb."

Such a comment would be an honest effort to reveal your negative emotions in order to invite the other members of the group to take a look at what's happening. This is not at all a destructive use of negative emotion. Compare the following outburst:

> "Hell, you people aren't getting anything else out of me. You're just a bunch of bumps on a log. A person could die in here, and you probably wouldn't even notice it. Someone spills his guts out in front of you, and you just sit and stare. I'll be damned if I'm going to say anything about myself in this group!"

In this second case, the person just dumps angry emotions on the rest of the group. Such dumping usually doesn't do very much to get the group to keep to its contract. It merely makes the others angry in return.

5. *Dealing with emotions as they arise instead of waiting.* It isn't very useful to save up your emotions and then dump them on others. If you save your emotions up, people won't even remember what caused them in the first place. It's much better to express them as they come up, even if they're negative. One of the goals of the group, after all, is to learn how to express emotions well. It's much easier for your fellow group members to handle your emotions as they come up instead of all at once. Consider the following example:

> *Marge* (in meeting 12): "Bill, I haven't really talked to you much at all over the past 12 weeks, but I feel I have been growing closer and closer to you. About the only way I can put it is that I love you."

All at once poor Bill is faced with Marge's saved-up emotions. Now he doesn't even know what to say. It's too much all at once. In

76

this case Marge hasn't been following the contract. She has done little to form and establish a relationship with Bill, and now she wants everything at once. She has set herself up for rejection by not finding out, from week to week, how Bill responds to her.

In this example, Marge had been storing up positive feelings about Bill. It may be even more frequent that we collect *negative* emotions about others. We save them up like trading stamps and then try to cash them in all at once. You're asked not to do this in the group. Rather, you're asked to try to deal with emotions as they come up, whether they are positive or negative. If you do have negative feelings and aren't getting them out, then at least say that. For instance:

> "I feel bottled up right now because of the way I feel, but I'm having a hard time expressing it."

Then others will help you get your feelings out where they can be faced.

There are constructive ways of getting at negative feelings. Take the following example:

> *Maria*: "John, you've been quiet for a long time this evening. This is making me nervous. When people are too quiet, I wonder what's going on inside. Sometimes I get to thinking that the quiet person is judging me, thinking that I'm making a mess of things or thinking that the group itself is no good. That's what makes me nervous. I've been asking myself this evening what it is that keeps you from jumping into the conversation."

It wouldn't have done Maria any good to wait a few meetings in order to tell John how she's feeling. By sharing with him her negative feelings as they take place, she avoids having to blow up at him. I don't mean to suggest that you should express any and every emotion as soon as it comes up. You have to be the judge of emotional experiences that need airing in the group.

6. *Controlling emotions instead of burying them.* It's more work to learn how to control emotions than it is to bury them, but the payoff is better. You may say to yourself "These are the days of freedom. What's all this about *controlling* emotion? Let it all hang out!" "Control," however, is a neutral word. The question is, control for what? The basketball player who practiced hard learning all sorts of

different moves could be spontaneous, because he or she could use whatever moves were needed during the game. Learning emotional control has a similar effect. If you're unafraid of your emotions, if you don't feel that you have to get rid of them or bury them because they just get in your way, if you can learn how to express many different emotions but *make them fit the situation*, then, emotionally speaking, you can be a spontaneous person. People who let all of their emotions hang out aren't spontaneous but rather undisciplined or uncontrolled. Their interpersonal lives are usually filled with fights and misunderstandings. On the other hand, people who exert too much control over their emotions are often seen as colorless and uninteresting. What is your interpersonal emotional life like? Is it too undisciplined? Too controlled? Or is it just what it should be?

7. *Sticking up for your emotional rights (emotional assertiveness)*. As we've seen, to be assertive means to stick up for your own rights without stepping on anyone else's rights or letting others step on yours. It means putting yourself forward without doing so at anyone else's expense. There are three possible ways of handling emotional situations: (a) nonassertively, (b) aggressively, and (c) assertively. Let's consider an example.

Suppose that one of the members of your group spends a lot of time saying nothing. He remains silent, just watching what the rest of you do. But every once in a while he becomes very active. During these active periods, he tells others how they aren't living up to the contract. When he's very active, he expects everyone else to be very active too. Now imagine that he's in the middle of one of his short active periods. He's challenging you because you don't start enough conversations with the other members. There are three possible ways in which you might handle your emotions in response to this person.

a. *Nonassertively—you just take it.* You apologize because you're not living up to the contract as well as you might. Inside you're as angry as you can be, but you try to swallow your emotion. You feel that he's being entirely unfair, but you sit back and take it. There's a big difference between what you feel and what you express. After the group meeting, you talk for a while with a friend and tell this person how angry you were in the group. You go home feeling that you've been used; you feel empty and still angry.

b. *Aggressively—you teach him a lesson.* You listen briefly to what he has to say, and then you let him have it. You challenge him angrily, telling him to practice what he preaches before trying to tell you how to run your life. You've been saving up your feelings about him for several sessions, and now you see a chance to dump them all.

78

There's no give-and-take; you just blast him. You immediately feel a little better, because you got a big load off your chest. When you go home, however, you wonder whether you've really accomplished anything. You don't like the idea of having to go back and face him. You feel a little guilty.

c. *Assertively—you let him know how you feel without punishing him.* First you listen to what he has to say. You may even tell him that there's some truth in what he says, that you deserve to be challenged on some of these issues. But you also tell him that it's hard to respond to him as well as you would like, because there are other issues getting in the way. You might go on to say something like this:

> "I know that a lot of the things you're telling me are true and that I should examine them in the group, but I have trouble listening to them from you. My feeling about you is that you're with us for short periods of time, but then you more or less disappear and say nothing. I've mentioned this to you before, telling you how it bothers me. Now I think that I resent your challenging me. I don't think that you've taken enough time to establish a relationship with me that would give you the right to challenge me. It's not that what you say is wrong. Frankly, I need this challenge, but I'd rather hear it from others in the group. I'd especially like to hear it from those who are closest to me."

This kind of direct, assertive behavior allows you to express your emotions without trying to destroy the other person. It also leaves the doors open for further communication. The question is not *whether* emotion—even negative emotion—is expressed, but rather *how* it is expressed.

Risk in Expressing Feelings and Emotions

There's always some risk in expressing emotion in interpersonal situations, because you're never quite sure how others will react. Nor are you absolutely sure that you'll stay in control or that what you're expressing emotionally is really the message you want to get across. Perhaps if you understand what makes it more of a risk to express emotions in interpersonal situations, you can get a better idea of how you want to shape your own interpersonal emotional life. In general, it's more of a risk to express your emotions *directly*. Let's see what this means concretely.

The difficulty-in-expressing-emotion chart. Take a look at the difficulty-in-expressing-emotion chart. The chart deals only with expressing feelings and emotions to or about people, not things. For instance, if I say "I can't get the lid off this jar, and I'm really getting angry," then my emotions are directed toward a thing rather than toward a person. Since we're dealing with interpersonal relationships, we'll discuss feelings and emotions that are directed toward people. The chart says that in many (if not most) cases it becomes more and more difficult to express feelings and emotions the farther down the chart you go. Therefore, it's relatively easy to express a positive feeling about a past situation when the feeling is directed toward a person who is absent. For instance, I may say this to you about George, who is out of town: "I really liked (positive feeling) the way George (who is absent) helped us decorate the hall for the dance (past situation)." At the other extreme is the expression of some negative feeling to a person who is present about something that the person is now doing. For instance, I might say to you "I find it annoying (negative feeling) to hear you (who are present) talking about Sally's faults (something you're doing right now)." This second kind of statement is ordinarily much more of a risk.

Feelings Difficult to Express

There are certain feelings that most people find quite difficult to express. As you examine your interpersonal style, try to identify which feelings you have difficulty expressing. Once you find out what they are, it's possible that you can work on expressing these feelings when they do come up. Some feelings that many of us find difficulty expressing are:

- *feelings of being no good.* "I just don't think I'm a very nice person. There's been a lot of garbage in my life, and at times I just feel like garbage."
- *feelings about not being able to do things.* "I'm a C student. No matter how hard I try, I don't do well in studies. Going to school does nothing for me. I feel like giving up."
- *feelings about not being able to do something about a situation.* "I can't seem to please you. When I'm silent, you let me know I'm not contributing. When I talk, you keep find-

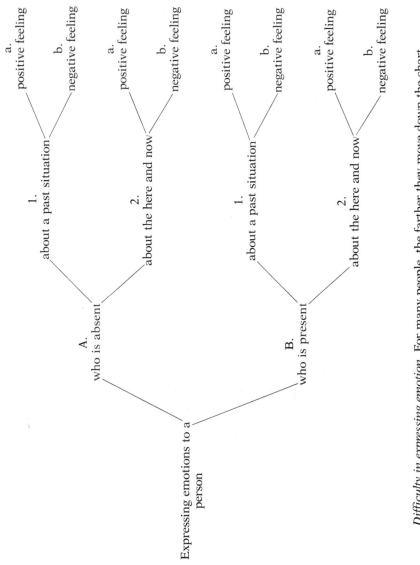

Difficulty in expressing emotion. For many people, the farther they move down the chart, the more difficult it is to express feelings and emotions. Therefore, the hardest would be to express negative emotions about a here-and-now situation to a person who is present.

ing fault with what I say. I just can't seem to live up to your standards. I don't know what to do, and I'm getting very angry."

- *not being able to handle affection from others.* "When you say you care about me, my whole body tightens up. I feel like running out of the room."
- *feelings of being hurt or rejected.* "You never talk to me in this group. It makes me think that you see me as dull or uninteresting. I go home feeling lousy and thinking about it."
- *feelings that come from a need to punish someone else.* "When you interrupt, Jane, I feel like calling you every name in the book. I could almost punch you."
- *feelings coming from a need to be punished.* "I want your attention so much I don't care how I get it. I even like it when you get angry with me and yell at me. I feel good even though what's happening to me is lousy."
- *feelings of guilt.* "I'm really depressed. Not only did I fail to help you when you asked me, but I lied to you. I almost feel like a criminal."
- *feelings of shame.* "I know I'm very homely. I even avoid looking in the mirror. I hate myself even more for making looks so important. I feel so embarrassed at times that it's like being raw inside."
- *feelings of being dependent on a person.* "I get frightened when I'm away from you for any length of time. I think I might need you too much. And that gets me down."
- *feelings of being passive.* "I know I don't do very much in this group. I sit around and wait for others to start or for others to begin talking to me. I don't really contribute anything. I feel useless, and then I begin feeling sorry for myself."
- *feelings of helplessness.* "When I talk here, nobody seems to notice me. It's awful to feel like you're talking to a brick wall."

This list could be lengthened. What feelings and emotions do you find difficult to express? In the group you're not asked to immediately express feelings and emotions you find difficult to handle. A first step is simply to let others know what emotions you have a hard time with. Learning how to take reasonable risks is an art in itself. The first rule is: move gradually; don't try to do everything at once. It's easier to express even risky emotions if there is a climate of trust in the group. Chapters 6, 7, and 8 talk about the kinds of skills needed to establish and develop an atmosphere of trust in the group.

Feelings about Yourself: Emotional Self-Awareness

All of us have certain feelings about ourselves. You may say such things as "I feel good about myself" or "I feel bad about myself," and no doubt these feelings often affect how you act. The way you think and feel about yourself is called your *self-concept*. Here are a few examples of how your self-concept can affect what you do.

- If you feel that you're no good, you'll hesitate to seek others to be your friends. After all, friendship means giving yourself to another person as a gift, and who wants to give someone a gift that's no good?
- If you feel bad about yourself, you'll also keep others who want to be friends with you at a distance because you're not "worthy."
- If you feel good about yourself, you'll make friends quite easily and let others be your friends.
- If you feel that you're a weak person who has lots of needs, you'll present yourself to others as a dependent person, as someone who needs a lot but can give little.
- If you feel that you're a strong person, you'll work at developing your talents and abilities and resources. You'll be active and won't wait around for others to do things for you.
- If you feel that you're a person who has lots of problems that you can't handle, you'll expect others to listen to your problems all the time. You'll expect them to become your helpers or counselors instead of your friends.
- If you feel that you're better than other people, then you'll have a hard time disclosing yourself to them, and you won't work at understanding them. You'll pay attention to some (the "better" or the "attractive" ones) and ignore others.
- If you feel that you're a boring person, you'll bring little enthusiasm to the group. You'll play the role of an uninteresting person and contribute little.
- If you feel that you're an energetic person, you'll bring this energy to the group. You'll be enthusiastic about learning, and you'll communicate your enthusiasm to others.

During the laboratory, then, it's important to examine your own feelings about yourself, because the way you think and feel about yourself—your self-concept—will influence the way you act in the group as well as your everyday interpersonal style.

Where does your self-concept come from? As you were growing up, the important adults in your life acted toward you in a variety of ways. If parents and others, such as teachers, paid little attention to you, you might have begun thinking deep down inside "I must be no good or uninteresting, because the important people in my life don't pay any attention to me." You might even have gone on to say to yourself (again, deep down inside) "Since I'm a dull, uninteresting person, it would probably be better to stay away from others. After all, who wants to spend time with a boring person?" In this way you may have become the kind of person who is always on the edge of the crowd, someone who never pushes himself or herself, a loner who is nevertheless quite lonely. Of course, the opposite may have happened to you. Perhaps the important adults in your life paid enough attention to you (though not *too* much) and showed you adequate care and love. Because of their attention to you, you were able to say deep down inside "Sure, I'm OK—at least as OK as the next person!" Your good feeling about yourself also affects your interpersonal style. You're not afraid to meet people, to make friends, to love and be loved, because basically you feel OK about yourself. If you learn some of the interpersonal skills and then find that you don't use them in the group (for instance, the skill of challenging), it may be that you don't feel good about yourself (you don't feel worthy enough to challenge anybody else). Part of your self-disclosure in the group should deal with how you feel about yourself. It might not be easy for you to examine how you feel about yourself, because it's sometimes painful to do so. My hope is that the trust and caring and support you receive from your fellow group members—your learning community—will help you examine your self-concept even when that's a bit painful to do.

A Checklist on the Appropriate Expression of Feelings and Emotions

This checklist can help you write your log and prepare your agenda for each meeting. Your ability to let emotions be part of your life has a great deal to do with your interpersonal style.

84

- Do I feel that it's all right to express my feelings and emotions and to talk about them? Or have I learned that feelings and emotions are dangerous?
- Do I allow my emotions to flow inside me spontaneously without trying to push them down? Do I feel that I have to manufacture or make up emotions that I don't really feel in order to please others or because I think I *should* feel certain emotions?
- Are there emotions that I overdo (perhaps anger, depression, self-pity)? Are there emotions I refuse to experience (attraction, hurt, enthusiasm)?
- Do I let my emotions add life and color to my conversations? Or do I control my emotions too much?
- Do I use my emotions to control others and get them to do what I want them to do (say, to leave me alone)?
- Do I express or talk about my emotions when they come up in the here and now, or do I save them for a later, safer time?
- Do I feel that I have a right to be emotional, to assert my emotions without stepping on the rights of others? Do I ever use my emotions to step on the rights of others?
- Am I willing to take reasonable risks in expressing my emotions? That is, do I experiment with emotions I ordinarily hide or avoid? For instance, can I, in a reasonable way, express negative emotions in the here and now to another person in the group?
- When others are expressing emotions, do I get scared? Do I try to get them to stop expressing their emotions?
- What emotions are most difficult for me to express? How can I get the other members of the group to help me express these difficult emotions?
- How do I feel about myself? Does the way I feel about myself prevent me from expressing my emotions or from even feeling emotions?

Who Is Responsible for Your Emotions?

Probably the most useful answer to the question of who is responsible for your emotions is that *you* are. Perhaps you can't control the external events that lead to your experiencing certain emo-

85

tions, but you can control the impact that these external events have on you.

Let's take an example. One of your friends fails to show up for a date, and you get angry. Now you can sit around feeling sorry for yourself and telling yourself "He has really made me angry." But notice who's suffering—it's you! You've given another person power over you to make you angry. If you recognize what's happening, you can, within limits, decide to change your emotional state. You can say to yourself "I'm responsible for my own emotions. For a while I did give Bob power over me to make me feel angry, but I'm going to set my anger aside and go about my business. I'll deal with the whole anger thing when I see him face to face." Once you take responsibility for what you're feeling, you have more of a chance to manage feelings and emotions that you find self-defeating. It isn't wrong to feel angry or depressed, but, if you keep blaming things or people out there, you end up hanging on to painful feelings. You end up choosing to be powerless.

The laboratory group is a good place to experiment with this new way of looking at feelings and emotions. Feelings and emotions will come up, but you can experiment with being responsible for whatever you're feeling, with *owning* your own feelings and emotions. For instance, if you're bored during a group meeting, you can say "I've just realized that I've given you people the power to make me bored. But I'm really responsible for my own boredom." You can see whether you then have a greater opportunity to do something about your boredom instead of sitting there feeling sorry for yourself. If you're angry with someone, one of the best ways to take responsibility for your own emotions is to deal with your anger with that other person: "I'm letting you get to me; I'm letting what you're doing make me angry. Now I'd like to talk to you about what I'm letting you do to me." Becoming responsible for your own emotions means making your emotions work for you instead of against you.

When emotions are expressed in terms of what another person does *to you*, then there is little you can do about your emotions, because you can't change the other person's behavior. However, if you express your emotions in terms of what's happening inside you, you can move to owning responsibility for your own emotions.

> *not:* "You don't accept me."
> *but:* "I don't feel accepted."
>
> *not:* "You don't seem to trust me."
> *but:* "I feel untrusted."

This distinction is important in making your communication less threatening to others. If you make others feel that *they* are responsible for your emotions, they'll hesitate to get close to you. "You don't accept me" is an accusation, whereas "I don't feel accepted" is a statement about yourself.

Emotional Self-Awareness: Exercises in Expressing Feelings and Emotions

These exercises don't ask you to feel things you don't really feel or to do things just to arouse emotions. Feelings and emotions will arise spontaneously during your give-and-take in the group. The exercises may help you to identify and express your emotions more fully and to discuss them more freely when you feel a need to do so.

Exercise 9
Learning Different Ways of Expressing Feelings and Emotions

Feelings and emotions can be expressed in many different ways. Sometimes we limit ourselves in the way we express emotions and thus rob our interactions of color. Feelings and emotions can be expressed:

- *by single words.*
  "I feel good."
  "I'm angry."
  "I feel trapped."
  "I feel awful."
  "I feel depressed."
  "I feel great."
- *by many different kinds of phrases.*
  "I'm in the dumps."
  "I'm sitting on top of the world."
  "I'm all messed up."
  "I'm in the dark."
  "I'm up against the wall."
- *by describing what's happening to you.*
  "I feel like I'm being dumped on."
  "I feel she loves me."

"I feel I'm being watched."
"I feel he cares."
"I feel they don't give a damn."
• *by describing what you'd like to do.*
"I feel like giving up."
"I feel like hugging you."
"I feel like telling them off."
"I feel like getting the hell out of here."
"I feel like singing and dancing."

The purpose of this exercise is to help you learn different ways of expressing feelings and emotions. Many different kinds of emotions are listed below. Try to express each of these emotions in four different ways. First look at the examples; then do your own.

1. *Joy*
   Use a single word: "I'm happy."
   Use a phrase: "I'm on top of the world."
   Use a sentence describing what's happening to you: "I feel that he loves me!"
   Use a sentence describing what you'd like to do: "I feel like taking a day off to celebrate!"

   Now do your own, using the same emotion—joy.

   Use a single word: _____

   Use a phrase: _____

   Use a sentence describing what's happening to you: _____

   _____

   Use a sentence describing what you'd like to do: _____

   _____

2. *Anger*
   Use a single word: "I'm mad."
   Use a phrase: "I'm ticked off."
   Use a sentence describing what's happening to you: "I'm being used by my best friend."
   Use a sentence describing what you'd like to do: "I feel like telling them off."

   Now do your own, using the same emotion—anger.

Use a single word: _____

Use a phrase: _____

Use a sentence describing what's happening to you: _____
_____

Use a sentence describing what you'd like to do: _____
_____

3. *Anxiety*
   Use a single word: "I'm nervous."
   Use a phrase: "I'm all up in the air."
   Use a sentence describing what's happening to you: "My stomach's doing flip-flops."
   Use a sentence describing what you'd like to do: "I feel like jumping out of my skin."

   Now do your own, using the same emotion—anxiety.

   Use a single word: _____

   Use a phrase: _____

   Use a sentence describing what's happening to you: _____
   _____

   Use a sentence describing what you'd like to do: _____
   _____

Now take the following emotions or emotional experiences and try to express them in a number of different ways. Remember that the purpose of this exercise is to loosen you up with respect to the ways in which you express feelings and emotions through words. See how colorful you can be.

- ashamed, embarrassed
- feeling defeated
- confused
- guilty
- rejected
- depressed
- peaceful
- frustrated
- pressured
- capable, competent
- feeling bad about yourself
- satisfied
- misused, abused
- without energy
- distressed
- loving
- bored
- hopeful

89

Exercise 10
What Happens to You When You
Experience Certain Feelings and Emotions

If you are to make sense out of the emotions of other people, you must first make sense out of your own. In this exercise you're asked to describe what you actually experience when you feel different emotions. In the last exercise you experimented with different ways of expressing common emotions. Now you're asked to look inside yourself and see what happens when you're feeling different emotions. What do you feel like? What happens to your body? What do you feel like doing?

Directions

a. First try to recall a specific time when you experienced each emotion.
b. Then try to describe as fully as possible what happened to you then and what usually happens to you when you experience that emotion.

Example 1: The feeling of being accepted

A time when I felt accepted: "Last week I told my girl friend about how lousy things were going in school. She listened to me and tried to understand me. She didn't tell me I was a lousy person. I felt accepted."

When I feel accepted:
    I feel warm inside.
    I feel safe.
    I feel I can be myself, whatever I am.
    I feel I can let my guard down.
    I feel like sharing myself.
    I feel my strengths more deeply.
    Some of my fears go away.
    I feel at home.
    I feel at peace.
    Some of my loneliness goes away.

Example 2: Feeling scared

A time when I felt scared: "I was riding in a bus with a bunch of other kids. The driver yelled 'God, the brakes are gone!' We finally stopped OK, but I was really scared until then. I thought it was all over."

When I feel scared:
I feel panic.
My mouth gets all dry.
I feel like going to the bathroom.
I feel like running away.
I feel like crying out.
I can't think straight.
I think only of myself.
I feel very weak and helpless.

Now try to picture yourself in situations in which you've actually experienced the emotions listed below. Briefly describe what the situation was like; then describe what happens to you when you feel that emotion.

1.	affectionate	2.	attracted	3.	competitive
4.	afraid	5.	bored	6.	defensive
7.	angry	8.	confused	9.	disappointed
10.	anxious	11.	that I belong	12.	free
13.	frustrated	14.	inferior	15.	ashamed
16.	guilty	17.	close to someone	18.	sad
19.	hopeful	20.	lonely	21.	satisfied
22.	hurt	23.	loving	24.	shy
25.	joyful	26.	rejected	27.	superior, proud
28.	jealous	29.	respected	30.	trusting

Are there some emotions you find hard to describe? Are there some emotions you don't experience? Are there some you avoid? Which emotions are hardest for you to experience? Which are the most painful?

Your instructor will make suggestions on how to share what you've discovered about yourself with your fellow group members.

Exercise 11
Recalling an Intense Emotional Experience

Take another look at the list of emotions in Exercise 10. Try to recall a few times when you felt one or more of these emotions in an intense or deep way. Try to recall emotional experiences that say something about the kind of person you are, especially the kind of person you are in interpersonal situations.

Example

"A friend whom I liked very much moved to another city. I didn't hear from him for over a year. Then one day I got a phone call. He said he was coming back for a visit. When I asked what was bringing him here, he said that he was really coming to see me. I was flabbergasted and excited. I couldn't believe what he was saying. I was so taken by surprise that I didn't know what to say and must have sounded dumb. He was going to be in town for a weekend. I got busy and made all sorts of plans. I was walking on air. I could hardly wait until he arrived. When he landed at the airport and I met him, I was so excited that I hugged him. I really wanted to be liked by him. We did a lot of things, like going out for dinner and talking far into the night. I don't know what I really wanted, but I kept thinking that more should be happening. I think that I was expecting too much. Our visit was enjoyable, but somehow it didn't meet my expectations. I was happy and disappointed at the same time. This confused me so much that at times I fell silent. I never talked about my confusion, so when he left I felt lonely, disappointed with myself, wondering what I really wanted, confused, and depressed. It made me think about the lack of closeness with others in my life. It made me wonder what I expect from relating with others. I'm not sure I know yet."

The purpose of this exercise isn't to make you reveal your secrets. If what you write the first time seems too personal to share with your fellow group members right now, try to write out one or two emotional experiences of some intensity that you feel you can share with the others. (If you have difficulty with this or any exercise, ask your instructor for suggestions.)

Exercise 12
Identifying Emotions That Are
Difficult for You

Some psychologists suggest that emotions we feel, even those we have difficulty in expressing, almost always get expressed. They say that, if we don't express them directly, we express them indirectly—perhaps through some *other* emotion that we find more acceptable. For instance, if I find it difficult to let you know directly that you've hurt me, I'll express it to you in some indirect way. I might get depressed. Or I might get angry with someone else I feel it's safer for me to get angry with. Or I might even try to be nicer and more affectionate with you; that is, I might try to express to you the *opposite* of what I feel. However, if I express my emotions indirectly, I confuse other people. Then I might complain that others don't understand me, when one of the reasons they don't is that I'm not straightforward with my emotions.

Directions

Review the list of emotions in Exercise 10 once more. Check those that are difficult for you to express or that are difficult for you to face when other people express them. Write out a short statement on what you do with difficult emotions, how you try to hide them, in what *indirect* ways you try to express them.

Example

"I find it difficult to express affection. I get very tense when I feel affectionate toward someone. I think I often feel guilty about having such an emotion. I'm very afraid to express affection in any physical way, because I'm afraid that others will reject me and think that I'm dumb. I guess I don't feel very attractive to other people, even to those I consider my closer friends. Instead of expressing affection directly, I do other things for people; I do favors for them. I try to find out what a certain person is feeling so that I can be considerate. I don't think this works very well at times, because the other person thinks that I'm taking care of him or her. The person feels mothered and then finds it hard to be with me. I get

93

depressed because I can't find ways of letting others know that I feel affectionate toward them."

In doing this exercise, try to deal with emotions that affect your interpersonal style. Also, recall the standards of appropriate self-disclosure. Try to write out some statements that you feel you can share with your fellow group members, if not now, perhaps later on.

Exercise 13
Feelings about Yourself

In this chapter I've talked about how you think and feel about yourself—your self-concept. Since your self-concept affects the ways you act with other people, this exercise is meant to help you examine the ways you feel about yourself. It should give you some insight into your interpersonal style.

Directions

Finish the following incomplete sentences. Don't spend a great deal of time thinking about what you *should* write. Write whatever comes first to your mind.

1. I get angry with myself when . . .
2. I like myself best when . . .
3. I feel ashamed when . . .
4. I trust myself when . . .
5. When I fail, I feel . . .
6. I feel on top of the world when I . . .
7. I get confused about myself when . . .
8. I'm pleased with myself when . . .
9. I get down on myself when . . .
10. I feel confident when . . .
11. When I do things that I feel are wrong, I . . .
12. When I'm successful, I . . .
13. I'm most at peace with myself when . . .
14. I get depressed when . . .
15. When I think of what others have told me about myself, I . . .
16. I get annoyed with myself when . . .
17. When I take a good look at myself, I . . .
18. I think I'm OK when . . .

19. I think I'm not OK when . . .
20. When I look at my values, I . . .

Now look at the way in which you've completed these sentences. What do they say about the ways you feel about yourself? Do you usually see yourself as OK? Or are you usually down on yourself? How do the ways you feel about yourself affect the ways in which you relate to other people?

Your instructor will give you further suggestions concerning how to use this exercise.

Concreteness: How to Avoid Being Vague in Communication

5

This chapter suggests ways of talking about yourself clearly and concretely. The skill of concreteness can make your conversations with others rich, crisp, and engaging. The three elements of concreteness–specific experiences, specific behaviors, and specific feelings and emotions–are explained and illustrated.

As you probably know, it's far too easy to talk without saying anything. It's very frustrating to try to listen to someone who is being vague or unclear, whether or not the person is being so intentionally. Vague talk creates distance between people who are trying to communicate with each other. Even though it's not always possible to be completely clear in your communication with others (because interpersonal communication can at times get very complicated), trying to be as clear as possible is certainly a desirable goal. People who play interpersonal games with one another use vagueness as part of the game. Have you ever come away from a conversation asking yourself "I wonder what he really meant?" Learning the skill of *concreteness*—the skill of being clear and specific in communication—is a way of avoiding even unintentional game playing.

Concreteness works two ways. First of all, in order to be clear, direct, and straightforward in your communication, you can demand concreteness of yourself. The appropriateness both of your self-disclosure and of your expression of feelings depends in part on your being specific and concrete. Second, you can ask others to be concrete in order to avoid misunderstanding and game playing. If another person speaks very vaguely concerning how he or she feels about you, you'll probably find it quite frustrating. But, as I said in the last chapter, you're responsible for your own emotions. In this case, being responsible for your own emotions means that you have a right to ask

97

the other person to clarify what he or she is saying about you. If you possess the skill of concreteness, you'll be able to help the other person sort out how he or she feels about you; that is, you'll be able to help the other person become more concrete.

Concreteness in Self-Disclosure and in the Expression of Feeling

Concreteness means being clear and specific instead of being vague and general. Notice the difference between the following two statements.

"Things aren't going too well."

This statement is so general that it could mean almost anything.

"I'm really frustrated. I want to go to a movie with John, but I have one of my sinus headaches."

This statement is concrete. It's clear and specific. You don't have to guess what's happening.

There are three areas concerning which your interpersonal communication should be concrete: experiences, behaviors, and feelings and emotions.

Experiences. By the word *experience* I mean what happens to you, including what other people do to you. Let's take a couple of examples.

"I have a bad headache."

This is an experience, since it's not something you're doing but rather something that's happening to you. A headache is something you're putting up with or enduring.

"Bill got mad at me and told me off when we were out last night."

Here what you describe is not something *you* did but rather something that someone else did *to* you. Therefore, it is an *experience.*

To be concrete when you talk about experiences, you must talk about clear and specific experiences. Suppose you make the following statement:

"Things are bothering me."

Here you're talking about an experience or a set of experiences—that is, you're talking about things that are happening to you—but it's not clear just what is happening to you. Now suppose you say this:

"My mother and father have been fighting a lot this past week, and I'm getting more and more upset."

Again, you're talking about experiences, but this time the experiences are clear and specific. *Concreteness means talking about clear and specific experiences happening in clear and specific situations.* If you had said "Oh, things are going on with mom and dad," you would not have been concrete, although this statement is less vague than "Things are bothering me." Let's take another set of examples.

"Things are OK right now."

This statement is a vague or general description of an experience or a set of experiences. It's not concrete.

"Sheila just gave me a massage, and I'm feeling very relaxed."

This, too, is an experience—a massage is something that is done to you—but it is a specific experience and is therefore concrete.
Experiences can also include thoughts and things that arise in our imaginations, especially when they pop up spontaneously without our willing them. For instance:

"That dirty remark she made about me keeps coming up in my mind over and over again."

When you talk about yourself in the group, the more concrete you are in describing your experiences, the more clear and straightforward you will be to others.

Behaviors. Your descriptions of your behaviors must also be clear and specific if your conversations are to be concrete. By the word *behaviors* I mean the things that *you do* rather than the things that happen to you or that others do to you. You're also talking about behavior if you mention what you *don't* do. Let's look at a few examples.

"I think I messed things up with my mom yesterday."

This statement is about behavior, but it's not concrete.

"I yelled at my mother yesterday for no good reason at all. I upset her, and she started to cry."

Here the behavior is much more clearly described. The statement would have been even more concrete if the whole scene had been described more fully, perhaps as follows:

"Yesterday, while I was eating lunch, my mother began to tell me about a party she'd gone to with her girl friends the night before. I told her that I didn't want to hear about her stupid party. I upset her quite a bit; she started to cry."

This is a long way from the statement, "I think I messed things up with my mom yesterday."

Thoughts and imaginings are also behaviors if you *deliberately* think certain thoughts or imagine certain things. For instance:

"All week I kept thinking about what you said to me in last week's group meeting. I kept thinking how right you were to point out the ways I've been defensive with you."

Another example:

"This week I tried to picture myself standing in front of my father and telling him that I want to move out of the house, that I want to be more independent."

These two statements describe behaviors, because in each case the thoughts or imaginings are deliberate.

Notice that there are experiences and behaviors that other people can see (*external* experiences and behaviors) and those that other people can't see (*internal* experiences and behaviors). Appropriate self-disclosure usually includes telling people concrete and specific things about both the external and the internal you.

Feelings and emotions. Concreteness also includes talking about specific feelings and emotions that take place in specific situations.

For example:

"This is the third time you've interrupted me this evening. It makes me wonder whether you really want to listen to me. I'm getting very annoyed."

Here feelings are expressed specifically, clearly, and directly. Interpersonal communication becomes vague and unclear when emotions are ignored or talked about too generally. For instance, you might say:

"What you say stirs up a lot of things in me."

Here you're telling somebody that you feel *something*, but it's not clear what your feelings are. They could be positive or negative feelings.

"What you just said makes me feel even more affectionate toward you."

This statement of emotion is specific and clear.

Actually, feelings and emotions are experiences, since they are things that happen to you (or, in keeping with the idea that you're responsible for your own emotions, they are things that you *let* happen to you). However, they are such special kinds of experiences that it makes sense to talk about them separately.

The Importance of Concreteness in Interpersonal Communication

Conversations that lack concreteness tend to be boring. If you find yourself becoming bored during the course of a group meeting, ask yourself whether you're being concrete in talking about yourself and whether the other members of the group are being concrete. The chances are that a good deal of the boredom comes from the fact that people are talking but not really saying anything. Besides making sure that, when you talk, you talk about specific experiences, behaviors, and feelings, there are other ways of making your conversations with others more direct, immediate, and concrete.

Use the pronoun "I." Your conversations in the group will be much more direct and concrete if you use the pronoun "I" when you are talking about yourself. This point might seem too obvious to mention; yet it's amazing how often people do talk about themselves without using the pronoun "I." If one of the members of the group says "People freeze up when they are challenged," when he really means something like "I froze up when you confronted me about being too quiet in the group," then his conversation loses directness and concreteness. As another example, notice the difference between saying "It's not easy to talk directly about yourself" and saying "I'm afraid to talk about my shyness here. I guess I'm afraid that someone here will hurt me."

There are many different substitutes for the personal pronoun "I": *"it's* not easy" for *"I* don't like," *"people* have a hard time" for *"I* find it difficult," *"you* have to really try hard" for *"I* don't try hard enough," *"we* are trying to understand you" for *"I* care about you and want to understand, but I'm having a hard time." Often we use these substitutes without even knowing it. But when we do use substitutes, we become less direct, we don't take responsibility for what we're saying or feeling, or we hide in the "we" of the group. You and the other group members can help one another with gentle reminders whenever someone starts using substitutes for "I."

Invite dialogue. Your conversation will probably stay much more direct and concrete if, to put it bluntly, you don't talk too long. It's much better to have conversations than speeches, whether in the group or in one-to-one interpersonal situations. The longer you talk, the less you say about yourself, and the less concrete you become. The way in which you talk to others should invite a response of some kind; it should invite dialogue. If you talk nonstop, nobody else can get a word in. Eventually people just stop listening, and you end up feeling ignored. One of your jobs in the group is to establish relationships. These relationships can be established only through the give-and-take of conversation. Talking in long, nonstop paragraphs keeps other people at a distance. In summary, therefore:

- Don't use too many words.
- Keep your sentences short.
- Speak from both your head and your gut.
- Expect the other person to reply.
- Give the other person a chance to reply.

Exercises in Concreteness

Exercise 14
Speaking Concretely about Your Experiences

In the following exercise you are asked to speak about some of your *experiences*—first vaguely, then concretely. Try to choose experiences that say something about your interpersonal style. In this way the exercise will also help you to make up your self-disclosure agenda.

Example 1

A vague or unclear statement of experience: "People don't always understand me very well."

The vague statement turned into a clear and concrete one: "People tell me that I come across as cold and aloof. As a result they find it hard to be friendly toward me. They tell me it's hard to talk to me." This statement also deals with the experience of how other people are reacting to the speaker, but now the experience is clear.

Example 2

A vague statement of experience: "People sometimes don't let others finish things in this group."

The vague statement turned into a clear and concrete one: "Jim, just now I thought I was saying something important about myself, and you interrupted me and started talking about yourself. I don't know whether what I was saying even mattered to you." The speaker is talking about a concrete experience that he or she is having right now in the laboratory group.

103

Now write out four examples of your own. Include first a vague statement about an experience of yours and then a concrete statement. In this exercise, stick to *experiences*—that is, to things that are happening or that have happened to you. Make each experience a real one, but also something you feel you can share with the members of your group.

Exercise 15
Speaking Concretely about Your Behavior

In the following exercise you are asked to speak about some of your *behaviors*. Speak about behaviors that relate to your interpersonal style either inside or outside the group.

Example

A vague statement about a behavior: "I'm not so good at challenging others."

The vague statement turned into a concrete one: "When I challenge others, I do so in a harsh way. I preach at them and tell them what they're doing wrong. I dump my own negative emotions on them, and this doesn't help them at all."

Now write out four examples of your own. First make a vague or unclear statement about your interpersonal behavior; then turn the vague statement into one that is clear and concrete. In this exercise stick to *behaviors*—that is, to things you do or fail to do.

Exercise 16
Speaking Concretely about Your Feelings

In this exercise you are asked to speak about your *feelings and emotions,* especially to the degree that they relate to your interpersonal style.

Example 1

A vague or unclear statement about feelings: "Sometimes I wonder about myself."

The vague statement turned into a clear and concrete one: "I often find myself feeling sorry for myself because I don't do as well in finding and keeping friends as others seem to do. I get down on myself and then just give up."

Example 2

A vague or unclear statement about feelings: "It seems that things are funny between you and me, Joan."

The vague statement turned into a concrete one: "Joan, when we talk, I let myself feel like a little kid. I let you frighten me just as my mother frightened me at times when I was actually a kid."

Now write out four examples of your own. Talk about feelings that are related to your interpersonal style or about actual feelings that come up when you talk with other members of your group. Try to say something about feelings that you find hard to express.

Exercise 17
Speaking Concretely about
Experiences, Behaviors, and
Feelings Together

In this exercise you are asked to put it all together; that is, you are asked to speak concretely about yourself, your interpersonal style, and what's going on with you in your training group.

Example

In this example one group member is talking about her relationship with another group member. First she makes a vague statement: "Things would be better if we related better." Then she turns this statement into an expanded, concrete statement—that is, one in which specific and clear experiences, behaviors, and feelings are talked about:

"I like you. That's my first reaction. You seem both smart and sensitive, and I like that combination. Usually when I like someone, I play it cool, hold back, find out whether I'm going to be liked back or not. But I feel like taking a chance with you and telling you right out how I feel. You seem interested in me. You start conversations with me, and, when I talk

105

about myself, you let me know that you understand. This is a chance for me, because I always fear that people I like will reject me. Right now I almost feel like I'm forcing you to like me."

Now write out three examples of turning a vague statement into a small, concrete paragraph that deals with your interpersonal style and/or your experience with your fellow group members. Make sure that your paragraph contains clear and specific experiences, behaviors, and feelings—all three. Be as specific and as real as possible. Imagine yourself actually talking to one of the group members or to some significant person outside the group.

Exercise 18
Speaking Concretely about
Your Personality Characteristics

The purpose of this exercise is to help you talk more concretely about yourself and your interactions with others. The exercise deals with those *personality characteristics* that are most likely to affect your interpersonal style. A personality characteristic is some habitual way that you act or are as a human being. For instance, if you said "I'm intelligent," you would be referring to a way you are and also, hopefully, to ways in which you behave (which show that you're intelligent). Some other examples are "I'm moody," "I'm a cold person," "I'm affectionate," "I'm overprotective," "I'm cautious," "I'm sociable," "I'm shy," "I'm generous," "I'm stingy," "I'm passionate," "I'm deep," "I'm grouchy," and so on. The trouble with personality characteristics is that, even though they do say something about you, they are not very concrete. Therefore, in discussing your interpersonal style, it isn't enough simply to say, for example, "I'm shy." This statement says something about you, but not very much. To examine your interpersonal style carefully, you must make your personality characteristics come alive by translating such statements as "I'm shy" into clear and specific experiences, behaviors, and emotions.

Directions

In this exercise you are asked to write down some personality characteristic that you think describes you or your interpersonal style and then to describe each characteristic more fully in concrete statements about specific experiences, behaviors, and feelings related to that characteristic.

Example

A statement of a personality characteristic: "I'm a cautious person."

A translation of this personality characteristic into concrete statements about experiences, feelings, and behaviors:

- "People have hurt me in the past."
- "I never speak my mind unless I'm absolutely sure I'm right."
- "I'm afraid that others will put me down if I speak up."
- "I feel guilty when I make mistakes, especially if my mistakes affect other people. So I don't risk making mistakes."
- "Because I've been hurt, I'm very fearful of hurting others. So I'm very hesitant to get close to people, even people I call my friends."
- "I make few friends. I expect people to come to me and make friends with me."
- "I think that I want a comfortable life. Putting myself on the line with others and getting close is sometimes painful. It makes me uncomfortable, and I don't like it."
- "I daydream a lot about being a smooth person in interpersonal relationships. Daydreaming makes me feel important, and there's no risk involved."
- "Some people have asked too much of me in a relationship—too much closeness. That bothered me."

These statements make up a much more complete picture of what it means for this person to say "I'm a cautious person."

Now write down four or five of your own personality traits and, in the same way, make them as concrete as possible by translating them into clear and specific experiences, behaviors, and feelings.

In all of these exercises your instructor will give you further suggestions concerning how to share what you discover about yourself.

The Skills of
Attending and Listening

6

So far we've examined interpersonal communication from the point of view of the person who is trying to communicate himself or herself to another person, and we've reviewed the skills necessary for effective self-disclosure. In this chapter and in the next one we turn our attention to the other side of interpersonal communication: the skills of attending and listening (Chapter 6) and the skill of responding with understanding (Chapter 7).

The Role of Attending and Listening in Communication

Communication is two-sided. If I choose to reveal myself in some way to you, I'll probably expect you to respond to me. If we are to involve ourselves in each other, we must have *dialogue* between us; our interactions must be marked by give-and-take. Just as there are skills of self-disclosure, so there are skills of responding. If I reveal myself to you, what do I expect in return? I expect that you will listen to me carefully and that you will let me know that you understand what I'm saying. There is in all of us a craving to be understood, at least by a few people. It's true that, if you reveal yourself to me, there are a number of different kinds of response I can give you besides understanding. For instance, I might tell you something about myself, or I might challenge what you've said to me, or I might ask you to tell me more about yourself. But these responses will probably make more sense to you if the *first* thing I try to do in responding to you is to understand you. Letting people know that I understand what they are saying to me is a kind of oil that lubricates the entire communication process.

If I am to let you know that I understand you, I must first *pay attention* to you and listen to what you have to say about yourself.

Knowing how to pay attention to others and how to listen carefully are essential communication skills. Poor attending and poor listening result in poor understanding. Since the quality of interpersonal communication depends on the quality of understanding that people show one another, it seems only right to start this part of our work introducing you to the skills that lead to good understanding—namely, the skills of attending and listening.

Helping one another without being helpers. The skills I'm going to discuss in the rest of this book, including the skills of responding, are also the skills that a person needs to be a good counselor or helper. Nevertheless, I don't mean to imply that you and your fellow group members are in the group in order to be counselors to one another. The purpose of training you in these skills is not to turn you into a counselor or helper in your interpersonal relationships either inside or outside the group. If you had a good friend who was a doctor, it would be annoying if he always talked to you as if you were his patient. He might do that at times when you needed it, but generally you would expect him to talk to you as any other friend would. The same thing could be said of a friend of yours who was a counselor. You wouldn't expect him or her to be counseling you all the time. If your friend did constantly act as a counselor toward you, you would have the right to resent it. On the other hand, he or she might be able to give you a bit of counseling when you needed it.

In the group, if you keep to the original contract and learn the skills presented here, you'll certainly *help* one another often enough, but you can help one another without taking on the role of counselors. Professional helping or counseling is a one-way relationship: the helper gives to the client. In the laboratory group, the skills you learn should help you to communicate better as *equals*. Establishing and developing relationships in the group is a two-way process, just as friendship is a two-way process that involves a lot of give-and-take. In the best of friendships, two people support and challenge each other without taking on the role of professional counselors with each other.

Social Intelligence: Knowing What to Do in Interpersonal Situations

As we saw in Chapter 2, each interpersonal skill has three parts: (a) awareness or perception—that is, knowing what's going on; (b) communication know-how, or the ability to translate your awareness into language that other people understand; and (c) assertiveness, or having the guts to speak up and deliver your messages. Learning the skills of paying attention and listening to others is the basic way of increasing or developing your awareness or perception (a).

Some people are intelligent in the sense that they are good in school and do very well at jobs in which school-type knowledge and skills are important. Other people, who may or may not do well at school tasks, are intelligent in the sense that they have a natural feeling for people. They know what to do in social and interpersonal situations. This kind of intelligence, which is not necessarily measured by ordinary IQ tests, is called *social intelligence*. When we say that someone has "street smarts," we mean, at least in part, that he or she is socially intelligent, that he or she knows what to do in people-oriented situations. If you are socially intelligent, you have good awareness or good perception in the following different ways.

• *You're in touch with yourself.* Being in touch with yourself means that you know what's going on inside yourself. You're able to tell how you're reacting when other people talk to you. For instance, as you listen to someone talk in the group, you might say to yourself "Eric is frustrating me right now because he isn't speaking concretely." You can also tell what reactions you have inside as you're about to talk to others. For instance, as you're about to talk to a good friend who has been cold to you for a while, you might say to yourself "I feel nervous right now, because I'm not sure why Carol has been cold to me." It's a skill to be in touch with what's going on inside you (especially your emotions) as you meet and talk with people. Some people are frequently out of touch with themselves. Have you ever asked a friend who is talking to you "What are you so angry about?" and received a surprised "Who's angry?" for an answer? In such a case, it's evident to you that the person has been speaking quite angrily, but he or she doesn't realize it. In other words, he or she isn't in touch with what's going on inside. People who are out of touch with themselves keep stumbling over themselves as they try to communicate with others.

Learning the skill of being in touch with yourself is an important part of this laboratory experience. One way to learn this skill is to get feedback from people in your group who have good awareness or perception.

• *You're in touch with what's going on inside others.* If you're a good listener, you pay close attention to another person and listen carefully enough to get an understanding of, a feeling for, what's going on in that person's "world." The world inside each of us is a

111

unique combination of feelings, thoughts, and imaginings. There is a science-fiction novel that portrays people in a future century as having the means to free themselves from their own bodies in such a way that they can swap their inner worlds. If you could accomplish this feat, you could float out of your own body and into not only the body but the mind and spirit of another person. You could directly feel what he or she feels, imagine what he or she imagines, and know what he or she knows. Of course, in the universe that we actually know, it's not so easy. Getting a view or an understanding of another person's world takes a great deal of work and involves a lot of attending and listening—all of which become second nature to the socially intelligent person.

• *You're in touch with what's happening between two people who are trying to communicate with each other.* This is a more complex skill. If you are using this skill, you're in touch with what's happening inside each person, but you're also attending and listening to the give-and-take *between* these two people. For instance, if Tom and Jean are talking, you're able to tell by listening to what each of them says whether they're taking the trouble to understand each other. Perhaps there comes a time in their talking with each other when you have to say "Either you two aren't listening to each other or you don't think that it's important to let each other know that you understand what the other is saying." Learning the skill of paying attention to and understanding the interactions between two people gives you the ability to enter conversations that others are having in a helpful way. If you don't have this skill, when you try to enter conversations you'll merely be interrupting what's going on.

• *You're in touch with what's happening in a group.* This is an even more complicated skill. In a group of only six people, since each person is relating to five other people, there are, in a sense, 30 different relationships going on. Moreover, what happens in one relationship often affects another relationship, which complicates things even more. For instance, suppose that in your group you develop a liking for both Carl and Sue. But then Carl and Sue have some kind of falling-out and remain angry and hostile for a while. Obviously their quarrel could affect and complicate your relationships with both of them. You might withdraw for a while, and stop interacting with either Carl or Sue until they straighten out their relationship with each other, because you don't want to be in the middle. Being aware of all the complications that are taking place in all of the relationships in a group and of how these complications affect other relationships is a difficult job.

112

• *You're in touch with the communication strengths and weaknesses of the people you meet.* In the group, having this skill means that, as you learn communication skills, you begin to perceive whether others are using these skills or not. You pay attention to *how* others communicate and begin to see *patterns* in the ways they communicate. Some of these patterns help communication, whereas others do not. For example, you might notice that Chris allows himself to get bored with group interaction but that he doesn't deal with his boredom immediately. Instead he saves it up and finally dumps it on the group in the form of anger. Chris hasn't learned the skill of taking responsibility for his own emotions, and his failure to learn it is getting in the way of his becoming a better communicator. By paying attention to how Chris interacts in the group, you get in touch with one of his communication weaknesses. Learning to become aware of the communication strengths and weaknesses of others is extremely important if you want to learn the skill of challenging others creatively. As we'll see, it's especially necessary (and often harder) to get an appreciation of the communication *strengths* of others. For some reason, it's much easier for us to see weaknesses. The problem is that merely confronting people with their weaknesses is a poor way of communicating with them. I'll discuss more fully the skill of paying attention to both weaknesses and especially strengths when I deal with the skills of challenging in Chapter 10.

It's evident that awareness or perception by itself isn't enough to make you a good communicator. You also need the communication know-how involved in each of the skills presented in this book. You've already been introduced to the know-how involved in letting others get to know you, and you've seen that the skill of self-awareness is an essential part of that know-how. Now it's just as important to learn how to become more keenly aware of others— what's going on inside them, how they communicate, and how they handle the communications of others.

The Skill of Attending: Being with the Other Person

At some of the more dramatic moments of life, just *being with* another person is extremely important. If a friend of yours is in the hospital, sometimes just your presence there can make a great difference, even if there can't be much conversation. Similarly, your being with someone who has lost a husband or wife or child or friend can be very comforting to that person, even if very little is said. Yet it's not only in dramatic moments that attending to or being with another person is important. On the contrary, attending to or being actively present to another person should be part and parcel of our everyday interpersonal-communication styles.

Perhaps you've suddenly exclaimed to someone during a conversation "You're not listening to me!" You noticed that the person was distracted, that he or she had drifted away someplace and was not paying attention. Such inattention is annoying, because it seems to reflect a lack of respect. It doesn't help if the other person says in self-defense "I am too! I can repeat every word you've said." You want more than the other person's ability to repeat what you say word for word; a tape recorder can do that. You want his or her *attention*. You want this person to *be with* you.

Physical Attending: Using Your Body to Communicate

Your body plays a large part in your interpersonal communication. What you do with your body can emphasize a message you're trying to communicate in words, or it can erase the message you give with your words and even substitute an opposite message. If you invite me to dinner and I turn to you, look you in the eye, smile broadly, and say with an enthusiastic voice "Great! When will we go?" then what I do with my body emphasizes my verbal message. Both my words and my body say that I really want to go out to dinner with you. However, if you ask me out to dinner and I turn away from you a bit, scratch my head, hesitate, look down at the ground, hunch my shoulders a little, and finally say "Yeah, I guess so" in a rather dull voice, my whole body is saying "no" even though my words are saying "yes." In this case, my body has the real message, and it's the opposite of the message my words are carrying.

The letters S-O-L-E-R can remind us of the five basic things we can do with our bodies to let others know that we're involved with them.

- S: *Face the other person SQUARELY.*
  This is the basic posture of involvement. If you face someone squarely, you say by your posture "I'm available to you; I choose to be with you." Turning your body away from another person while you talk to him or her lessens your involvement with that person. Even if you're seated in a group, you can turn in some way toward

114

the person you're speaking to. By directing your body toward the other person, you say *"I am with you right now."*

- *O: Adopt an OPEN posture.*
Crossed arms and crossed legs can be signs of lessened involvement with others. An open posture—especially open arms—is a sign that you're open to the other person and to what he or she has to say. An open posture is a nondefensive position.

- *L: LEAN toward the other.*
This is another sign of presence, being-with, availability, involvement. Watch two people sometime in a restaurant booth who are really engaged in conversation. Very often they're both leaning forward as a natural sign of their involvement. Then look at people who are leaning back and looking around the restaurant. They seem to be bored. They certainly don't look involved.

- *E: Maintain good EYE contact.*
As you speak with another person, spend much of the time looking directly at him or her. Some people object and say that such eye contact is unnatural; they see it as staring the other person down and think that it makes people feel uncomfortable. However, the two people who are deep in conversation in the restaurant booth have almost constant eye contact with each other. It doesn't look unnatural at all, because they *are* deeply involved with each other.

- *R: Try to be at home or relatively RELAXED while attending.*
If you're really involved with others and want to communicate with them, you can be both intense and, in a sense, relaxed at the same time. You're relaxed because you're doing something natural. If you're attending to another person but you're on edge and rigid, then you're not at home with this person, and perhaps it would be best to deal openly with your anxiety. If you're taking an oral exam from a teacher, you may be attending to him or her but still be very much on edge. But if you're the same way in talking with a friend, you should begin to ask yourself "What's going on here? What's making me so nervous?"

Your body is always communicating. It's impossible not to communicate. What we say, what we do, how we say it, how we do it, what our bodies are like during all of this, all add up to constant communication. The suggestions represented by S-O-L-E-R aren't meant to be rigid rules that you should never violate. Rather, they are norms or standards for direct communication that you can use to ask

115

yourself what you're doing and saying with your body. Your body communicates whether you want it to or not, and it's therefore important to learn what your body is saying at any given moment. It's even better to learn how to *use* your body—your bodily position, your posture, your gestures—to say what you want to say. If you don't use your body to communicate, it will go on communicating anyway, perhaps giving others messages that contradict the message you're giving through your words. Perhaps you've never thought much about the fact that your body sends messages, just as your words send messages. If so, you may feel a little self-conscious and awkward at first as you try to discover what your body is saying in your conversations.

What you do with your body is important not only in one-to-one conversations but also in the group. Some people think that, since there are a number of people in the group, they can take time out from time to time. That is, they think that they can sit back or even slouch, meanwhile letting their minds wander, and that the group will not be disturbed at all. Experience shows that this belief is simply false. The person speaking is often disturbed by the person or persons who are not attending, even though he or she is receiving a great deal of attention on the part of one or two. A nonattending member also affects the rest of the group. Very often three or four attending members (because they *are* attending) notice that one or two people have pulled back from the group either physically or psychologically. Perhaps they don't say anything about what they notice, but they are distracted by such nonattending all the same.

Psychological Attending: Active Listening

I've discussed physical attending in two ways: first in terms of your being aware of your own body and of how it communicates and second in terms of using your body to pay attention to and to be with others. Therefore, physical attending has a purpose of its own; it's a form of communication in its own right. However, another reason for paying attention to others is to *listen* to them. On the surface, listening seems easy. It almost seems to be a passive behavior that we can't help but engage in. But there is a significant difference between mere hearing and *active listening*. So far attending has been described as a physical activity; active listening is the psychological activity involved in attending.

116

Total listening means that, as you attend to another person, you try to "listen" to all the messages that he or she is sending. Messages can be sent many different ways—by talking, for instance, but also by remaining silent. Total listening means paying attention to the three sources of the other person's messages: the person's nonverbal behavior, the person's voice, and the person's words.

Listening to nonverbal behavior. The face and body of the person speaking to you are extremely communicative. Even when two people are silent with each other, the atmosphere can be filled with messages. Moreover, psychologists have found that, if your facial expression gives a different message from your words, most people will believe the message they find in your face. This is especially true of simple emotional messages. If I'm attending to you, and your words say "I like you and want to be with you," while the message in your face says "I'm not too sure that I like you, and I both want to be with you and don't want to be with you," I will believe your nonverbal rather than your verbal behavior. Since the body "talks," it's important that you learn how to tell what the body of the other person is saying.

Nonverbal behavior punctuates verbal messages in much the same way that periods, question marks, exclamation points, and underlinings punctuate written language. Nonverbal behavior can punctuate or modify interpersonal communication in at least the following four ways.

• Nonverbal behavior can *confirm* what is being said. If I say to you that I agree with what you're telling me and nod my head at the same time, my nonverbal behavior confirms what I'm saying. If you smile when you tell me that you appreciate the way I'm trying to understand you, your smile tells me that you're sincere about what you're saying to me. When nonverbal language is saying the same thing as verbal language, nonverbal language is confirming.

• Nonverbal behavior can *deny* or *confuse* what is being said. If I say to you that I'm not upset by your challenging me to be more active in the group, but my upper lip is quivering as I say this to you, my nonverbal behavior seems to deny the message of my words. Or if you say that you're angry with me and smile while you're saying it, your nonverbal behavior confuses me. An amazing number of people take back serious messages by smiling while they deliver them. The smile seems to say "I'm uncomfortable, either for myself or for you, in saying this to you."

117

• Nonverbal behavior can *strengthen* or *emphasize* what is being said. If, as I say "I don't want to go now," I pound my fist on the table, my behavior emphasizes my desire not to go. It indicates how *intensely* I feel about it. Or if I throw my arms around you as I say "Thanks for helping me yesterday," my hugging you adds to the intensity of my expression of gratitude.

• Nonverbal behavior can add *emotional color* to what is being said. Emotional color is another form of emphasis. If, when you tell me that you don't want to see me any more, I say to you that I don't like hearing that and then bury my face in my hands and cry, it's obvious that I'm deeply hurt by your rejection of me. Or if, after telling you that you've just confronted or challenged me without first trying to understand me, I stare at you silently with a frown on my face and my eyes half-closed, my nonverbal behaviors tell you that I'm angry with you. If I proceed to stalk out of the room, my nonverbal behavior adds even greater intensity to my display of anger. Or if I tell you that I'll go along with your plans but frown and shake my head from side to side as I do so, I'm telling you nonverbally that I'm quite reluctant to go along with you.

Listening to the quality of the speaker's voice. Messages are sent out not only by a person's body and facial expressions, but also by the way the person uses his or her voice. The *way* in which words are said can punctuate or modify verbal behavior in the same ways that nonverbal behavior can.

• Voice quality can *confirm* what is being said. If you tell me that you want me to challenge you when you begin talking too much in the group, and if you tell me this in a firm, steady, clear voice, then the quality of your voice confirms your verbal message. If I tell you that I'm upset, and if my voice is nervous and quaking, my voice confirms what my words are saying.

• Voice quality can *deny* or *confuse* what is being said. If I ask you to go to a movie and you reply in a very slow, halting voice "Well . . . sure . . . I think I'm free . . ." the way you're using your voice denies what your words are saying. The real message is not in the words but in the *way* you delivered the message. If I'm smart, I'll suggest that we go some other time. Or if you tell me that you like me, and if then you laugh a little, I'll probably be more preoccupied with the meaning of your laugh than with what you said. Delivering the message "I like you" with a laugh is confusing. The message is no longer clear.

• Voice quality can *strengthen* or *emphasize* what is being said.

118

If I say to you "Get out, and stay out" in a deliberate way and in a harsh tone of voice, the way I use my voice says that I really mean what I'm saying. My voice *underlines* my words. Or if I say that I want to help you work out more concrete agendas for the group when you're writing your log and say so in a deliberate, clear, and sincere way, my voice emphasizes my verbal message.

    • Voice quality can add *emotional color* to what is being said. If you say in a soft, halting, broken voice that you didn't mean to hurt my feelings, your voice adds the emotions of grief and shame to what you're saying. Or if I tell you that I like you in a voice that seems almost ready to sing, then the quality of my voice adds joy and enthusiasm to my words.

Total Listening: Putting Nonverbal Behavior, Voice, and Words Together

In a way, we've been looking at the process of human communication in bits and pieces. A low-level communicator often fastens on this bit or that piece. For instance, he or she notices a person's nonverbal behavior but fails to put it together with the person's words. A high-level communicator, on the other hand, is a total listener. He or she listens to *all* the cues and messages sent out by the other person—nonverbal behavior *and* voice quality *and* words—in order to see how they either confirm and complement one another or contradict one another. Just as a person with good academic intelligence sees the relationship among a number of different ideas or concepts, a person with good social intelligence sees the relationship among a number of different kinds of interpersonal-communication behaviors.

A high-level communicator also has a feeling for the *context* in which communication is taking place. By *context* I mean the situation or the setting, including the emotional setting, of the communication. A high-level communicator attends not only to nonverbal, voice, and verbal messages but also to the *background* of these messages. Let's take an example. I see you making certain feeble efforts to pay attention during a group meeting. You talk about yourself, but you don't sound enthusiastic. You take time out and pull back physically more often than usual. When you do make contact with another group member, you seem to be doing so out of a sense of duty. However, if I know that you've just had a fight with your parents, I can read all of your interpersonal-communication behavior *in the context* of your fight with your parents. In this way I'm listening not only to your nonverbal behavior, the quality of your voice, and your words themselves but also to the *situation* or *context* in which you find yourself. Understanding the context or background helps me to understand your verbal and nonverbal messages better.

It takes skill and practice to be able to attend in such a way that you can put the total communication package together. Attending, then, together with listening, is an *active* process. Attending also means, however, avoiding two extremes. On one hand, it means that you're not passive. If you're passive, you'll miss many of the messages the other person sends through nonverbal behavior, voice, and words, and you'll miss entirely the context from which the person is speaking. On the other hand, attending means that you're not so active a listener that you read *too much* into the messages the other person is giving. Reading too much into messages is playing a game called "psychologist." A person who plays the psychologist game makes up messages and meanings that don't really exist.

Listening for Experiences, Behaviors, and Feelings

It's a skill to be able to identify the feelings that are related to another person's experiences and behaviors. Often people don't come right out and say "I'm happy," or "I'm sad," or "I feel affectionate toward you." Their feelings and emotions are found in their words, their voice quality, and their nonverbal behavior. You have to use these sources to *identify* feelings if you are eventually to let the other person know that you *understand* his or her feelings. For instance, suppose that you received the following note from a friend of yours:

> "Why didn't you show up for my birthday party? Was it too much to expect one of my best friends to come? I kept waiting for you to walk in the door, and you never did! I didn't even get a phone call from you! What's this all about?"

Of course, in a note or a letter you don't get either the voice quality or the nonverbal behavior of the writer but only the words. And notice that your friend doesn't say explicitly how he or she feels. All the same, can you tell how your friend feels?

He or she feels _____.

Some possibilities are: hurt, angry, confused, ticked off, mad, very annoyed, put out, disappointed. Notice that you don't have to prove that your friend had any of these emotions when he or she was writing the note. The emotions almost leap out at you from the page. If he or she were talking to you in person, it would be even easier to see what emotions were being expressed, because the quality of voice (tight, high pitched, yelling) and the nonverbal behavior (frowning face, clenched teeth, a shrug of the shoulders) would carry much of the emotional message.

Identifying the source of feelings and emotions. As you listen to a person speak or as you read his or her words, there are basically two things you should attend to: the feelings being expressed and the source of these feelings (that is, the *experiences* and the *behaviors* that give rise to or cause these feelings). Consider the following example. One of your fellow group members says:

> "I don't know whether I should really continue in this group experience. Almost everybody seems to pick up these skills faster than I do. Maybe I'm not meant to have much of an interpersonal style. I think I try. Can there be such a thing as trying too hard?"

The person who says this is sitting with slumped shoulders and looking at the floor. Her voice is slow and low-pitched. What does she feel?

She feels _____.

Some possibilities are: down, dejected, depressed, like giving up, like throwing in the towel, miserable, sorry for herself. But now we ask: What is making her feel this way? What experiences or behaviors underlie these feelings? To practice getting at the sources of feelings, we'll use expressions of the form "She (he) feels _____ because _____." Now take what the girl has said, and try to complete the following:

She feels like giving up because _____

_____.

Some possibilities are: She feels like giving up because:

- she sees herself not doing as well as others.
- she's trying hard and not making the progress she wants.
- learning interpersonal skills is hard, slow work for her.

In summary, once you've identified the feeling or feelings that a person is experiencing, you should connect them to the experiences and/or the behaviors that underlie these feelings. Good attending behavior, including the ability to identify feelings and the experiences and behaviors that underlie feelings, is the foundation of *understanding*, which means appreciating someone else's world, seeing things from another person's point of view. As we've already seen, being attentive to another person is a powerful behavior in itself. It makes the other person feel appreciated and respected. It can be very rewarding to feel that another person is "with you" and paying attention to you, especially if you feel that the person is sincere.

This chapter assembles all the communication elements you need in order to be accurate in your understanding of others. The next chapter deals with the actual skill of *communicating* your understanding to others.

Exercises in Attending and Listening:
Increasing Your Awareness

Your instructor will tell you which of the following exercises to do. They all have one purpose: to help you become a more perceptive person.

Exercise 19
The Messages of Body and Voice

This exercise gives you practice in "reading" both nonverbal and voice-quality messages. However, since similar nonverbal and voice behaviors can indicate more than one kind of message, the exercise also encourages you to be *cautious* in interpreting nonverbal behavior. For instance, what message do you get when you see someone moving his or her head from side to side? This gesture might show reluctance, doubt, hesitancy. If your father is giving you money that you haven't earned because you need it for a date, he might say, while shaking his head, "Well, here it is, but I'm not very comfortable with giving it to you." His shaking his head is a sign of his reluctance. On the other hand, if someone is looking at you with an affectionate smile and shaking his or her head from side to side, the moving of the head can mean something like "I don't believe you; you're really too good to be true." In this case the behavior is playful and the message completely different.

Directions

Below is a list of nonverbal or voice-related behaviors. Try to give two or three meanings to each behavior.

1. A person nods his head up and down.
2. A person turns her head rapidly in a certain direction.
3. A person smiles slightly.
4. A person's lower lip quivers slightly.
5. A person speaks in a loud, harsh voice.
6. A person speaks in a low, monotonous voice.
7. A person suddenly opens his eyes wide.
8. A person keeps her eyes lowered as she speaks to someone else.
9. A person speaks in a very halting or hesitant voice.
10. A person yawns.
11. A person shrugs his shoulders.
12. A person is sitting very rigid and upright in her chair.
13. A person has his arms folded very tightly across his chest.
14. A person wrings her hands.
15. A person holds his chair tightly with his hands.
16. A person's breathing is quite irregular.
17. A person starts to turn pale.
18. A person keeps fiddling with his shirt collar.
19. A person slouches in her chair.
20. A person is constantly squirming.
21. A person inhales quickly.
22. A person continuously moves her legs backward and forward.
23. A person hits his forehead with his hand.

24. A person digs her heels into the floor.
25. A person who ordinarily doesn't stutter begins to stutter.

Share what you've written with your fellow group members. See how many different interpretations there are for each behavior.

Exercise 20
Eyes Closed: The Importance of Nonverbal
Messages

The purpose of this exercise is to give you some idea of how much you depend on nonverbal messages or cues in order to understand fully what another person is saying.

Directions

Choose a partner from among your fellow group members. Sit facing your partner. *Close your eyes.* Have a three-minute conversation on some topic related to the goals of this laboratory experience; in other words, talk about your interpersonal style with your partner. Keep your eyes closed during the entire conversation.

Once the three minutes are up, open your eyes and tell your partner how you felt talking to him or her without being able to see any gestures or facial expressions. What nonverbal cues did you miss the most?

Exercise 21
Giving Feedback on Nonverbal and
Voice-Related Communication Style

This exercise helps you to become more aware of the nonverbal and voice-related dimensions of other people's communication styles.

Directions

The group should divide into groups of four (A, B, C, D). Members A and B spend about five minutes talking to each other about a topic related to the goals of the group (interpersonal style, their feelings about what's happening to them in the group, or some other related topic). Member C observes

A, and D observes B. While A and B are talking to each other, C and D each jot down what they notice about the nonverbal and voice-related behavior of the person he or she is watching. After five minutes or so, the conversation stops, and C and D give feedback to A and B. Here are some samples of the kind of feedback that could be given.

- "Most of the time you spoke very quickly, in spurts. It gave me a feeling of nervousness. I think you do that a lot."
- "You sat very still during the whole conversation. Your hands remained folded in your lap the whole time. You looked somewhat rigid or like a very well-behaved kid. It gave me the impression that you were uncomfortable."
- "You tapped your left foot almost constantly."
- "When you began talking about how easy it is for you to get hurt by others, your voice stumbled a bit and you looked away from your partner. It gave me the feeling that this is a very touchy topic for you."
- "You put your hand up to your mouth quite a bit. It made me feel that you might be unsure of yourself."
- "You didn't have very good eye contact when you were talking about yourself, but you did when you were listening to your partner. I began to wonder how you feel about yourself when you talk about yourself."
- "You were *so* relaxed through the whole conversation— you even slouched down a bit—that I wondered whether you were really interested in what you were doing."
- "You used your hands to gesture a lot. It added a lot of energy to what you were saying. You looked so alive and involved."

After C and D give their feedback, the roles are switched. A and B watch the conversation and after five minutes give the same kind of feedback to C and D.

Remember that the purpose of this exercise is to make you aware of nonverbal and voice-related messages, both your own and those of other people. Care should be taken, however, not to read too much into any one gesture or fluctuation of the voice. This kind of overinterpretation can lead to the psychologist game. For instance, suppose I said to you "You cocked your head to one side during the whole conversation. It made you look confident in yourself, but I think it really meant that you were uncomfortable and not very confident in yourself." Here I would be giving too much interpretation. In doing this exercise you should (a) observe nonverbal behavior and voice-related cues; (b) report what you see to your partner; and (c) add how this kind of behavior affected you. Leave any further interpretation alone.

Exercise 22
"Right Now I Am Aware That . . . ."

The purpose of this exercise is to help you become more aware both of yourself and of another person as the two of you are talking together. You are asked to be especially aware of your own and the other person's nonverbal and voice-related communication behavior. The exercise should also make you more aware of what goes on inside you as you talk with another person.

Directions

Choose a partner from among your fellow group members. Sit facing each other. Take turns saying *single sentences* to each other, each time beginning with the words "Right now I am aware that . . . ." Finish the sentence by making some observations about the nonverbal or voice-related behavior of your partner, *or* about your own nonverbal and voice-related behavior, *or* about what is happening inside you that your partner cannot observe. Spend five minutes or so in this conversation, going back and forth one sentence at a time.

Example

A partial conversation might look something like the following.

A: "Right now I am aware that you are leaning toward me."
B: "Right now I am aware that you have a smile on your face."
A: "Right now I am aware that I'm holding tightly to the side of my chair."
B: "Right now I am aware that my mouth is a bit dry and that my heart is pounding a bit."

A: "Right now I am aware that your voice is steady even though you say your heart is pounding."

B: "Right now I am aware that I'm glad that you're my partner."

A: "Right now I am aware that what you just said catches me by surprise."

B: "Right now I am aware that my stomach has just tightened up a bit."

A: "Right now I am aware that your voice is soft and gentle."

B: "Right now I am aware that we have begun to talk more personally than I thought we would."

A: "Right now I am aware that I am a bit more relaxed, that I'm not holding on like crazy to the side of this chair."

And the conversation moves wherever the two people want it to.

After the conversation is over, tell each other how you felt during it and what you might have learned about yourself and your communication style. Did the two of you get any closer even though the exercise is artificial? How aware are you of what happens inside you when you're talking with another person?

Exercise 23
Experimenting with Physical Attending
in a One-to-One Conversation

The purpose of this exercise is to help you realize how physical attending affects a conversation.

Directions

Choose a partner for a one-to-one conversation. Have a conversation related to the goals of this laboratory experience. However, attend physically only as you are directed by your instructor. He or she will ask you to take various attending or nonattending positions. After the conversation, discuss how it feels to attend to someone who isn't attending to you or how it feels to have a conversation in which neither of you is attending. Your instructor will give you further directions for this exercise and help you discuss it afterward.

Exercise 24
*Experimenting with Attending in a Small
Group*

The purpose of this exercise is to help you realize the importance of attending in a group. Your instructor will give you
specific directions. Larger classes should break down into
groups of six or eight. You'll be asked to assume various degrees of attending and nonattending and afterward to talk
about how you felt about the attending or nonattending of
others and what demands intense attending places on you in
a group conversation. Your conversations in the small groups
should be about goal-related topics, such as interpersonal
style, how you feel as a member of this laboratory experience,
what's happening to you as you try to learn interpersonal-
communication skills, and so on.

Exercise 25
Paying Attention to Feelings

I've talked a great deal about paying attention to nonverbal
and voice-related behaviors in addition to the actual words of
the person you're speaking with. This kind of total listening is
needed in order to learn what the other person is *feeling* about
what he or she experiences or does. The purpose of this exercise, then, is to help you become sensitive to the feelings of the
other person. In the next chapter you'll learn the skill of letting the other person know that you understand what he or
she is feeling.

Directions

Read each of the following statements. Then write down a
number of adjectives or phrases that accurately identify what
the person is *feeling*. It's possible to identify feelings by reading the speaker's statement (after all, that's what we do when
we read letters). However, since we learn a great deal about
feelings from a speaker's nonverbal and voice-related behaviors, your instructor may read the following statements
aloud. In trying to pick out the feelings behind the following
statements, try to imagine each statement being spoken by
one of the members of your group.

Example

"Hey, things are going too fast here! I think that people are revealing too much about themselves too soon. I'm beginning to feel that I have to say things about myself that I'm not ready to say yet. Let's slow down!"

How does this person feel? pressured, nervous, under the gun, anxious, scared, disturbed, panicked, like putting on the brakes

Now complete the following.

1. *Rob speaks to Stephanie:* "I'm finding it hard to relate to you. You don't talk to me very much. When you do talk to me, you usually say what I'm doing wrong in the group."

   How does this person feel? _____

   _____

2. *Kevin addresses the whole group:* "This group is one of the first real learning experiences of my life! The skills I'm learning I can use in all of my relationships outside the group. I never thought it would be so practical!"

   How does this person feel? _____

   _____

3. *Bill speaks to Joan:* "Joan, I didn't mean to come on like a ton of bricks. It's clear to me now. I've been pushing you for the last three meetings, but I haven't really tried at all to find out how you've been feeling about me. I hope you won't give up on me."

   How does this person feel? _____

   _____

4. *Brian speaks up during a rather long group silence:* "Boy! These silences! When I was a kid, the family used to eat meals in silence, and I hated it. I mean, don't we have a lot to say to one another? My bet is that we have too much to say!"

How does this person feel? _____

_____

5. *Susan talks to Carl:* "Boy, am I glad that you finally talked to me. I've been wondering whether I did something to you or hurt you. I've been sitting here wondering how to talk to you. But now I find out that you haven't been having bad feelings about me at all!"

   How does this person feel? _____

   _____

6. *John talks to Don:* "Don, I've been thinking about you a lot this past week, and, well, I don't know what to say. You're a very sensitive person; you pick up on what I'm feeling very well. But even though you give to me, you don't seem to expect to receive anything from me. You let Mary give you support, but I'm not sure whether you want any support from me."

   How does this person feel? _____

   _____

7. *Tom speaks to Laura:* "Laura, you and I seem to have a lot of fun together. When we're having fun, it's really easy for us to talk to each other. But, at least as I see it, when we start talking seriously to each other, something happens. I feel myself pulling away. I was thinking about this during the week, and I didn't even know whether to bring it up or not. I keep wondering whether I want you deeper into my life than you are—or whether you want to be. I want a really solid relationship with you, but then I wonder what *you* want."

   How does this person feel? _____

   _____

8. *George speaks to Carl:* "I don't know how to say this to you. I appreciate the attention you show me because I like you. You let me know in little ways that you're with me, and yet you also let me know that you're my equal, not my helper or my parent. This increases my liking for you because I see you respecting me and I know I respect

you. I'd like to be able to meet some of your needs without being a helper or a parent to you."

How does this person feel? _____

_____

9. *Nancy talks to the group:* "I'm learning something in this group that I don't know what to do with. I've never thought about whether I like myself or not. Now I'm not sure I do. Between group meetings I compare myself with others who are doing so much better than I am. I know I shouldn't run away, but . . . what do I do?"

How does this person feel? _____

_____

10. *Kathy speaks to Maureen:* "Now that we've yelled at each other a bit and gotten the feelings out in the open, it doesn't seem too bad to me. This is so much better than having all those feelings blocked up and being afraid to say anything about them. It's funny, I think I now see you more clearly, and I see myself more clearly. How can that be? After our fight things are somehow much better than before."

How does this person feel? _____

_____

There are some suggested responses for this exercise in the Appendix. You can check what you've written with the responses given there. Your instructor will tell you how to share your responses with your fellow group members.

Exercise 26
Identifying the Experiences and Behaviors
That Cause Feelings

In order to communicate understanding to another person, it's essential to identify not only feelings but also the experiences and behaviors that give rise to these feelings. Your perception or awareness of another person is full or complete only if you can identify both the person's feelings and their source.

Directions

In this exercise, you are asked to review the ten statements in Exercise 25 and this time to identify the experiences and/or behaviors that seem to be causing the feelings expressed by the speaker. First note how this is done in the example that follows; then complete the other ten.

Example (see Example in Exercise 25)

This person feels <u>anxious and pressured</u> because <u>he doesn't want to disclose too much about himself too soon.</u>

Now complete the other ten.

1. Rob feels <u>hurt and rejected</u> because _____

   _____.

2. Kevin feels <u>pleasantly surprised</u> because _____

   _____.

3. Bill feels <u>ashamed</u> because _____

   _____.

4. Brian feels <u>disturbed and on edge</u> because _____

   _____.

5. Susan feels <u>relieved</u> because _____

   _____.

6. John feels <u>confused and hurt</u> because _____

   _____.

7. Tom feels <u>in a bind</u> because _____

   _____.

8. George feels <u>appreciative and loving</u> because _____

   _____.

9. Nancy feels <u>depressed</u> because _____

   _____.

10. Kathy feels at peace because _____

_____.

There are some suggested completions to this exercise in the Appendix. Compare what you've written with what you find there. Your instructor will give you directions on sharing your responses with the other members of the group.

The Skill of Responding with Understanding

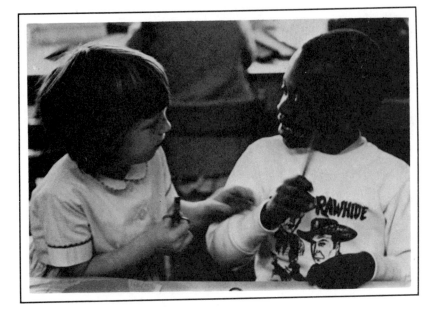

This chapter deals with a skill that is at the very heart of the process of interpersonal communication. The basic message is: if you want to communicate effectively with others, you must become effective first and foremost in the skill of responding with accurate understanding.

Introduction

More than any other communication skill, the skill of responding with understanding helps to create a climate of support and trust between two people or among the members of a group. Let's begin our discussion of this skill with an example. Suppose that you make the following statement to me about yourself:

> "I'm dreadfully shy. I feel that, if people get to know me, they won't like me. I've been this way ever since I can remember. I stay by myself a lot, but that doesn't solve my problem; and anyway I get terribly lonely. I'm here in this group as a kind of last resort. I don't know what else to do."

There are three basic ways in which I could respond to what you're saying about yourself: I could evaluate your statement, I could listen without responding, or I could respond with understanding.

1. *Evaluating: Our first instinct.* It's amazing how often our first instinct is to respond to others with some kind of evaluation of what they're saying. We do this so often that we don't even notice it. For example, suppose you say the following to me about yourself:

> "I don't have the guts it takes to get through school. I'm a very, very average person. Every time I fail an exam, I just

135

want to kick myself. Everybody's down on me—parents, teachers, the people I study with. Sometimes I think they don't give me a chance."

There are a number of ways in which I could respond to you by evaluating or judging either you or what you're saying about yourself. Here are some examples.

- "Hey, there's nothing wrong with you!" (I judge you to be good and evaluate your statement as not having any validity.)
- "That's a dumb way of acting. It's so self-defeating." (I evaluate your statement as correct and then judge you to be stupid for thinking, feeling, and acting in the ways you do.)
- "What are you going to do about it?" (I judge your statement as correct and then tell you that it's time to act.)
- "I'm not sure that that's the real problem. You've been kicked around a lot in life—for instance, by your parents. You've got to wash them out of your system." (I evaluate what you have to say about yourself by interpreting your statement and reading a lot into it.)
- "A lot of people go through this same thing." (By this cliché I judge that you don't really have a problem.)
- "Let's go to a movie tonight and forget all about it." (I judge that you're not saying anything worthwhile, and I indicate that by changing the subject.)
- "Try standing up for your own rights. Yell at other people once in a while, and they'll begin to take you seriously. This will shape them up and give you breathing room." (I evaluate you to be right in your judgment of yourself and then go on to give you advice.)

Whenever we evaluate others, we *decide* whether they're right or wrong. This kind of deciding—whether or not it's as open and clear as in the preceding examples—goes on constantly in most conversations. Listen to the conversations around you sometime with this in mind. It seems to be practically an instinct to evaluate, to judge, to see what a person is saying as right or wrong and to see the person as good or bad. Yet the process of evaluating and judging doesn't pull people closer together; it sets them farther apart. Often enough it leads to arguing. It sets one person up as better (more knowledgeable, more effective, more perceptive, more intelligent) than another. None of these consequences leads to effective interpersonal communication.

2. *Hollow listening: Listening without responding.* Often, when people say "I'm a good listener," what their statement means is that:

- they don't respond by evaluating or judging what others have to say;
- they pay attention to the person speaking and they look involved;
- they indicate their involvement by nodding their heads, by good physical attending, by saying such things as "uh-huh" and "yeah";
- they sometimes respond with such clichés as "I understand" or "I think I understand."

Such listening, however, is hollow if it consists merely of listening and nothing more. People in groups know this instinctively when they say to someone "I can see that you're with me, but you don't say anything." *The ultimate proof of good listening is good responding.* A person who says that he or she is a good listener but who in fact doesn't respond very much is often a person who finds it difficult to be active and to get close in interpersonal relationships.

3. *Responding with understanding.* Responding with understanding is perhaps the most useful yet least used response in interpersonal communication. Too often we let our instincts get the better of us and respond by evaluating rather than understanding. Responding with understanding means that:

- you listen carefully to the other person's total communication—words, nonverbal messages, voice-related cues;
- you try to identify the feelings the person is expressing and the experiences and behaviors that give rise to these feelings;
- you try to communicate to this person an understanding of what he or she seems to be feeling and of the sources of these feelings;
- you respond not by evaluating what he or she has to say but by showing your understanding of the other person's world from that person's point of view.

137

Let's return to the statement about shyness in the first paragraph of this chapter:

> *You:* "I'm dreadfully shy. I feel that, if people get to know me, they won't like me. I've been this way ever since I can remember. I stay by myself a lot, but that doesn't solve my problem; and anyway I get terribly lonely. I'm here in this group as a kind of last resort. I don't know what else to do."

Active listening *plus* responding without evaluating or judging might produce a response something like this:

> "It really gets you down to keep thinking that other people are going to reject you. This group may or may not help, but you don't know what else to do."

Let's see what this response is *not*:

- It is not an evaluation.
- It is not a judgment.
- It is not an interpretation.
- It is not a challenge.
- It is not advice.
- It is not just a word-for-word repetition of what you've said.

What this response *is* is an attempt to communicate an understanding of what you're going through. It's an attempt to give you a feeling of *being understood*. When someone reveals a problem, it isn't necessary to solve the problem or to come up with some kind of answer. Understanding of the person who has the problem is much more useful. Let's take another example. Sheila says:

> "I've just met somebody I really like. Our values seem to be a lot alike. And he's very willing to talk about his values. He's caring and even affectionate. But he's also a doer; he has goals, and he works hard. We've talked a great deal. I've really been looking for a person like this. It's still early, but I hope we can develop a solid friendship."

Consider the following three responses.

- *Evaluating.* "Be careful. You've been too enthusiastic in the beginning. If you rush in too quickly, you might get hurt."

  Now write out two other *evaluating* responses to what Sheila has said.

  a._____

  _____

  b._____

  _____

  What effect do you think such evaluating or judging responses will have on Sheila?

- *Hollow listening.* "Uh-huh. Yeah, that sounds OK." How do you feel when someone listens to you but says practically nothing?

- *Responding with understanding.* "He seems to have a lot of the qualities you look for in a good friend. It sounds like it could be exciting."

  Now write a statement expressing understanding, but use your own words.

  _____

  _____

Responding with understanding is not the whole of the interpersonal-communication process, but it is the basic ingredient in that process. Communicating understanding and being understood are basic to building trust. To communicate understanding to others is to provide them with the kind of support they need to reveal themselves more deeply and involve themselves more closely with you. I will be emphasizing the skill of responding with understanding throughout the rest of this book, because, first, so much of human relating depends on it. For instance, if you want to disclose yourself to a friend, you'll probably do so only if you think that he or she will understand you. Or if you want to challenge a friend, you can do so effectively only if you first take pains to understand him or her. A second reason for emphasizing this skill so much is that it is so unfor-

139

tunately rare in human dialogue. As you read this chapter and as you begin practicing this skill with your fellow group members, you are asked to judge its usefulness for creating a climate of trust and support in your own interpersonal relationships.

Like all communication skills, the skill of communicating understanding has three parts:

- *Awareness.* Chapter 6 dealt with the kind of awareness you need to be able to communicate understanding accurately. If you listen to me totally, you can get to know me "from the inside."
- *Communication know-how.* The rest of this chapter deals with how to communicate understanding to others.
- *Assertiveness.* Others won't know that you're "with" them unless you actually talk to them with understanding.

Responding with Understanding: Communication Know-How

Let's start our discussion of communication know-how and understanding with the following example. One of your fellow group members is getting a lot of advice from the others. Finally he says:

> "Hey, just a minute here! What the hell is going on? Do this, do that, do the other thing! Where's all this advice coming from? You people don't even know me, and you're telling me what to do!"

In order to be able to communicate understanding to this person, you would first have to be aware of two things: (1) what the speaker is feeling ("he's pretty angry right now . . .") and (2) the experiences and/or behaviors that are the source of his feelings (" . . . because everybody in the group is challenging him, but no one is making an effort to understand what's going on inside him"). Once you've successfully identified what the speaker is feeling and the experiences and/or behaviors that give rise to these feelings, the next step is to put your awareness or perception into words:

> "*You feel* pretty angry right now *because* everybody is ganging up on you, but no one is taking the time just to understand what you're trying to say."

140

So, good perception is necessary for good communication, but it certainly isn't enough. In groups too many members tend to keep their perceptions locked up inside themselves, where they do little good. In order to emphasize both (a) the two elements that must first be perceived accurately (feelings and what underlies feelings) and (b) communication know-how, we'll use the following formula for responding with accurate understanding:

"You feel_____(feeling word or phrase)_____
because___(experiences/behaviors that underlie the feeling)___."

Using this formula isn't the only way of responding with understanding, but for the present it can serve to remind you of the elements that are necessary for good responding. Later on in this chapter you'll be asked to drop the formula and to use your own natural way of speaking. For instance:

Formula: "You feel angry because, even though you study hard, you still don't get A's in your psychology courses."

Natural: "You feel that you're studying hard in your psych courses, but you still don't get A's. And that really ticks you off."

Of course, I'm not saying that the formula "You feel _____ because _____ " is never natural. You can continue to use it whenever it seems natural *to you*.

Responding with Understanding of Feelings

In this section I'll concentrate on the first element involved in responding with understanding—*feelings*. In Chapter 4 you were asked in a variety of ways to become aware of, identify, and give expression to your own feelings. If you can do that well, then you can also learn how to become aware of, identify, and express understanding of the feelings of others. In Chapter 4 you also saw that there are a number of different ways of expressing feelings, such as:

- by a single word: "I feel depressed."
- by a phrase: "I feel down in the dumps."

- by the word *like* with both experiences and behaviors: "I feel like the world has collapsed on me"; "I feel like giving up."

Any of these different ways of expressing emotion can also be used to respond with understanding of someone else's feelings. Let's look at an example. A friend you haven't seen for a year or two tells you that she has just graduated from nursing school. Just before graduation she had seen the director of nursing at a nearby hospital and had been promised a job in the intensive-care unit. This hospital is the one she has always wanted to work in. Now she has just heard that she won't get the job, but she hasn't received any explanation why. You can respond with understanding of her feelings in the three different ways I've described:

- by a single word: "You must feel bewildered" or "You must feel terribly disappointed."
- by a phrase: "You must feel up in the air" or "You must feel in a quandary."
- by the word *like*: "You must feel like storming in there and finding out what's going on."

Your responses will show more accurate understanding if you keep in mind that there's a big difference between *thinking* and *feeling*. Consider the following example.

"I *think that* I got a raw deal from my math teacher."

This is a judgment about the math teacher's fairness. Notice the difference in the next statement:

"I *feel angry* because I got a raw deal from my math teacher."

Anger isn't a thought or a judgment. It's a feeling. Sometimes we use the word *feel* when we really mean *think*. For example:

"I feel that I'm putting a lot of effort into learning these communication skills."

Notice the difference between this statement and the following one:

"I *know* that I'm putting a lot of effort into learning these

142

communication skills, and I *feel great* because I'm doing a good job."

In communicating understanding to others, it's important not to confuse thinking and feeling. Suppose that John says:

"I can tell that people don't like to spend much time with me. They think I'm boring or something like that. Well, I guess that's their privilege. I can't do anything about that."

In the following reply, thinking is kept distinct from feeling:

"John, you *think* that others don't like you very much, and that makes you *feel* down."

Exercise 27
Responding with Understanding of Feelings

In this exercise, communicate understanding only of the *feelings* of the speaker. Express this understanding in two different ways, using the formula "You feel _____" with a single word ("angry"), a phrase ("in the dumps"), or a "like" expression ("like giving up").

Example

Stan says to Maria and Mike:
"When you two talk about yourselves to each other, you do it so easily. I can watch you, but I don't seem to know how to join in. You seem so comfortable, while I sit here like a lump, not able to say a thing. I just don't contribute anything."

a. "You feel incompetent (awkward, inferior, useless)   ."
b. "You feel out of place (all tied up, down on yourself)   ."

Now respond in two different ways to each of the following. Show an understanding of the feelings expressed.

1. "This is a hell of a mess! Everybody in this group is ready to talk, but nobody seems ready to listen. Are we

all so self-centered that we can't take time out to listen to one another?"

    a. "You feel _____."

    b. "You feel _____."

2. "You and I have been fighting each other for weeks and not listening to each other. We compete. I think that in this meeting we did what we've been afraid to do. We talked to each other. We listened to each other. And you know, it's been very good talking *with* you rather than *at* you."

    a. "You feel _____."

    b. "You feel _____."

3. "Otis, when you talk directly and caringly to me, I find you a breath of fresh air! I thought you'd be the last one to challenge me. You're the first one who has told me how self-centered I can act in a way that I could hear."

    a. "You feel _____."

    b. "You feel _____."

4. "We've been talking about how Len and I relate or don't relate for almost a half hour now. How much longer does this have to go on? Let's put it on the shelf for a while. I mean really!"

    a. "You feel _____."

    b. "You feel _____."

5. "I want to get more involved in this group. At least I think I do. When I talk about myself, I know I'm not very concrete. I'm not sure that I'm really trying—or even that I want to try."

    a. "You feel _____."

    b. "You feel _____."

6. "I'm not sure that I can trust you. At times you're very outgoing and lively here. At other times you withdraw into yourself. I don't know what to make of you. I'm not sure I can count on you."

    a. "You feel _____."

    b. "You feel _____."

7. "Listen, Chuck, I've already admitted that I have a hard time getting in touch with my feelings. You keep asking me how I feel. I hate that question! What do you want from me?"

   a. "You feel _____."
   b. "You feel _____."

8. "Don't ask me any more about how I relate to my parents. I don't want to talk about it any more. I feel that I've already told you too much."

   a. "You feel _____."
   b. "You feel _____."

9. "I paid my money for this course, and I'm going to get something out of it. And nobody here's going to stop me. You can do what you want. But I've got goals here."

   a. "You feel _____."
   b. "You feel _____."

10. "I was up all night with my youngest child. He just wouldn't stop crying. And then my mother-in-law decided that this was the day that we should talk things out. Well, I'm here tonight anyhow, and I'm going to put all that aside for now."

    a. "You feel _____."
    b. "You feel _____."

Check your responses with the responses suggested in the Appendix. Were your responses or the responses of some of your fellow group members better than the ones in the Appendix?

Responding with Understanding of What Underlies a Person's Feelings (Content)

To understand a person, you must communicate understanding of more than just feelings. The experiences and the behaviors that give rise to a person's feelings are the "content" of understanding.

Therefore, to respond with full understanding to another person, you must respond both to his or her feelings and to the content (the experiences and behaviors that cause the feelings). Very often a person won't tell you directly how he or she is feeling (although you may get that information from nonverbal or vocal cues), but the person will tell you what he or she *thinks* underlies the feelings. Remember, you're trying to understand the other person from his or her point of view. Therefore, it's important to communicate an understanding of the situation *as the other person sees it*, even though you might think that he or she isn't seeing the whole picture. Your first job is always to communicate to the other person that you can see things through his or her eyes without evaluating or judging what you hear.

> *A friend:* "I don't find any kind of challenge in school. The courses don't offer much. The instructors are really just plastic people. I sit around class twiddling my thumbs. I'm not really motivated to do much outside class."

As you listen to a statement like this one, what is your first impulse? How would you like to respond?

- "Boy, you're always complaining. You don't take any responsibility."
- "You said it! I'm bored to tears in this second-rate place."

Both of these responses are evaluative and judgmental. The first passes judgment on the speaker; the second, in complicity with the speaker, passes judgment on the institution. The kind of response recommended in this chapter would merely show an understanding of your friend's feelings and of what gives rise to those feelings:

> "It's all very boring for you. School, teachers, courses—none of these seems to have much to offer."

This response recognizes your friend's feelings (boredom) and what he or she sees as giving rise to the feelings (school, teachers, courses that aren't challenging). Notice that this kind of response doesn't take sides. It doesn't suggest that either your friend or the school is to blame or not to blame.

Exercise 28
*Responding with Understanding of What
Underlies the Other Person's Feelings
(Content)*

The purpose of responding to feelings and to the sources of feelings is to communicate understanding, not to "psych out" the other person. Therefore, when I talk about understanding what "underlies" the other person's feelings, I don't mean coming up with deep, buried psychological reasons for the feelings the other person is experiencing. Rather, I mean communicating to the other person that you see that person's world from his or her point of view.

Directions

This exercise is really a continuation of Exercise 27, so you will be responding to the same ten statements found in that exercise. In order to help you focus on just what *underlies* the feelings expressed in the statements, I've provided feeling words or phrases in the formula and now ask you to supply the "because" part of the response.

1. "You feel almost like ditching this whole group experience because _____
   _____."

2. "You feel like you're finally getting somewhere because_
   _____."

3. "You feel pleasantly surprised because _____
   _____."

4. "You feel fed up because_____
   _____."

5. "You feel hesitant because_____
   _____."

6. "You feel hesitant about getting in deeper because__
   _____."

7. "You feel up against the wall because_____
   _____."

147

8. "You feel uneasy because_____
   _____."

9. "You feel like pushing ahead because_____
   _____."

10. "You feel played out because_____
    _____."

Check your responses with those in the Appendix and with those of your fellow group members.

Exercise 29
The Full Communication of Understanding:
Both Feelings and Content

I've been using the word *content* to mean what underlies a person's feelings (the experiences and/or behaviors that cause the feelings). In this exercise you are asked, first, to put feeling and content together in your response, using the complete formula "You feel _____ because _____." Then you are asked to put the formula aside and use your own way of expressing understanding. The formula is good for learning and sometimes works even in day-to-day conversations. However, if you use it too frequently, you'll come across as awkward or phony. Other people will think that you're using them to practice on. A high-level communicator picks up the *spirit* of understanding, knowing that it's important to communicate to others an understanding of what they're thinking, feeling, and saying in the flow of conversation. A high-level communicator ultimately communicates instinctively, weaving a great deal of understanding into ongoing conversations in a way that isn't self-conscious. A low-level communicator, on the other hand, always sounds like a person practicing a skill that has been poorly learned. Part of social intelligence is to be natural in the communication of understanding as well as in the use of every other interpersonal skill.

Directions

Read the following statements. Try to imagine that the person is speaking directly to you. You have two tasks:

a. Respond with accurate understanding, using the formula "You feel _____ because _____."

b. Write a response that includes understanding of both feelings and content, but use your own language and style. In this second response, try to be as natural as possible.

Example

Group Member A: "I had a hard time coming back here today. I felt that I had talked about myself pretty deeply last week, and I got angry when I thought that no one was taking the time to understand me. This morning I was trying to think of excuses I could use for not coming here tonight."

a. *Group Member B* (using formula): "You feel awkward and uneasy about being in the group tonight, because you got hurt and angry last week and you're not sure how all of this is going to be handled tonight."

b. *Group Member B* (using own style): "It's not easy for you to be here tonight. It sounds like the anger and hurt from last week haven't been resolved."

Now complete the following.

1. "John, why do you have to compare me to Jessie and Sue? I do that so much myself—always trying to be someone else or to live up to someone else's standards. It's something I'm really trying to break myself of. And then you come along and compare me, too."

   a. _____

   _____

   b. _____

   _____

2. "Gary, you seem to have everything so together. You're good at all these skills we're trying to learn. You even seem strong when you're talking about your weaknesses. I begin to compare myself with you, and all I can see are my own inadequacies."

   a. _____

   _____

b. _____

_____

3. "One thing that I'm beginning to learn here is that I have to stand on my own two feet. I never thought I could do it, but learning a few of these skills has changed me a lot. I can be assertive when I want to be, and that feels good."

   a. _____

   _____

   b. _____

   _____

4. "Jane, when I talk to you, I get goose bumps. I've put you up on a pedestal, and now I don't know how to deal with you. Each group meeting I say that I'm going to talk to you, but up until now I've chickened out every time."

   a. _____

   _____

   b. _____

   _____

5. "George, you haven't called me for a week. I know you don't have any obligation to call me, but I guess I have to admit that I want you to. It may be dumb, but I sit there waiting for the phone to ring—and it doesn't."

   a. _____

   _____

   b. _____

   _____

6. "I get the feeling that you're asking me to share with you how I feel about myself as a sexual person. You've shared yourself, and I think that you expect the same from me. I'm not sure that I feel I have to do that, unless it's what I want to do and the time is right for me."

a. _____

_____

b. _____

_____

7. "Last week you said that it was all right with you if we didn't go to the lake for a week. But now you seem to be depressed and on edge. I'm not sure whether how you feel has to do with not going to the lake or with something else. I'm wondering whether I've done something wrong."

a. _____

_____

b. _____

_____

8. "I know that I should never have started seeing Steve outside the group. I knew that sooner or later others would know and maybe even think that we were dealing with the real issues outside. Or perhaps talking about the other members behind their backs. Well, I guess that I have to face it all now."

a. _____

_____

b. _____

_____

9. "All of us in this group have said again and again that we want to get closer to one another. It's beginning to sound like a broken record. But none of us seems willing to risk very much. And I'm as much to blame as anyone else."

a. _____

_____

b. _____

_____

10. "I've been having a lot of trouble in school this semester. I think that I'm going to fail a couple of subjects. What's worse is that it's messing up my relationships with my friends. People begin to avoid me when they see me messed up."

   a. _____

   _____

   b. _____

   _____

Check the Appendix for some suggested responses.

Helpful Mutuality

The primary reason for teaching you the skills of self-presentation (Chapters 3, 4, and 5) is not to help you explore your problems as if you were a client in a counseling situation but to help you share yourself appropriately with others as you establish and develop relationships. These skills help you to become more responsibly intimate with others. In the same way, your primary goal in learning the skill of communicating understanding is not to become a counselor (although it's true that effective counselors must be very good at this skill) but rather to improve the quality of your communication in interpersonal relationships. Intimacy is fostered by *mutual* self-sharing and *mutual* communication of understanding. Certainly, good, solid dialogue between friends, even though it isn't the same as helping or counseling, can be very helpful. To put it another way, at times there are striking similarities between good counseling and high-level communication between friends. As we've seen, however, there is at least one major difference: in counseling one person is the helper or counselor, while the other is the client, whereas in a good friendship each friend both gives and receives help. There's something wrong with a friendship in which one person always seems to be the helper and the other the client. The give-and-take of friendship is what I mean by the term *mutuality*.

In this group experience, the goal is mutuality. If your group gets divided into helpers and clients, you'll be engaged in group counseling rather than in human-relations training. However, if there is genuine mutuality (give-and-take) in your group, you and your fellow group members will also receive a great deal of help from one another. For example, you may help one another change whatever you don't like about your present interpersonal styles. Let's take a look at a sample dialogue in which the participants help one another without taking on the roles of helper and client. For the purposes of this example, we'll focus on three members of a training group: Ann, Bill, and Chuck.

Ann: "We've been in the group about seven weeks now, and I've learned something very important about myself. I've just begun to realize how easily I get hurt. For instance—and I know that this will sound crazy—the fact that you [she looks at Bill] haven't talked to me yet this evening is bothering me. I'm beginning to see how overly sensitive I am, both here in the group and with my friends outside."

Bill (to Ann): "And so my not talking to you when you expect it of me—well, that hurts. But now it's making you take a look at just how sensitive you are. We make quite a pair! I'm beginning to see that I'm not sensitive *enough* about other people's feelings. And I don't seem to get hurt very easily. But I think I don't get hurt because I keep my relationships superficial and bland. I don't let people get close enough to me to hurt me. And that's a hell of a way to live!"

Ann (to Bill): "So you feel that you lack sensitivity. In one way it makes your life more comfortable—you have fewer problems with people like me. But you don't like your lack of involvement, because it makes life too bland. You're right; we do make quite a pair."

Chuck (to both Ann and Bill): "This clears up something for me. At times, Ann, I've felt that you were rather cold to Bill. I haven't said anything because Bill hasn't reacted very much, even when you've answered his questions in a very uninterested tone of voice. Now that you, Ann, talk about being overly sensitive, and you, Bill, talk about not being sensitive enough, your interactions, such as they have been, make sense to me. I feel that I can talk easily to both of you. I like both of you, and I think that you like me. You've both taken time to understand what I've had to say about myself, and I like that."

Ann: "Bill, we both do get along well with Chuck, but these differences in our interpersonal style have kept the two of us from spending time trying to understand each other. I think we can probably do something about that. I want to say to everybody in the group that I'd like to explore the ways in which my being too sensitive are affecting my relationships with all of you. I'd like to get as concrete as possible."

Chuck (to Ann): "I feel closed off from a couple of people here, too. I'd like to explore why it's so easy for me to relate to some and so much harder to relate to others. What's happened between you and Bill gives me a little more courage to put this on my agenda."

In this conversation no one becomes a client, and no one assumes the role of counselor. Ann, Bill, and Chuck try to understand one another, and they are all willing to explore areas of their interpersonal style that need looking into and perhaps improvement. If the same pattern of communication continues with them and with the other members of the group, they'll undoubtedly get a great deal of help without becoming helpers and clients.

Exercise 30
Responding with Understanding in
Everyday Life

To make responding with understanding part of your natural communication style, you'll have to go beyond written exercises and even beyond doing it in your training group. The whole reason for the written exercises and the group experience is to enable you to communicate understanding freely and naturally in all of your conversations—with family, with friends, with fellow students, with people at work. The following simple procedure might help you increase your communication of understanding.

1. Begin to become aware of conversations between people from the viewpoint of accurate understanding. Do people

frequently convey accurate understanding? See how often people respond to others with a judging or evaluating statement.

2. Observe yourself in your ordinary conversations. Notice how often you convey accurate understanding naturally as part of your communication style. At first, don't try to increase the number of times you use the skill of responding with understanding outside the group. Merely observe your usual behavior.

3. Begin to increase the number of times you use the skill of responding with understanding in daily conversations. Don't go overboard; just begin to use it somewhat more frequently than you ordinarily do. You want to be genuine, not phony, in your use of this skill.

4. Begin to use the skill of responding with understanding as often as you think it's called for. You'll probably discover that there are a great number of opportunities for using this skill. For instance, when someone confronts you or merely yells at you, try to respond first with accurate understanding of the person's feelings ("My coming late has really made you angry") before responding to the confrontation itself ("Dad asked me to run to the store just before I came over here, but I should have called and told you that I'd be late"). You'll probably find that the communication of accurate understanding is an excellent lubricant that tends to make interactions go more smoothly.

5. Try to observe the impact that your increased communication of accurate understanding has on others and on your interpersonal style generally. Once you naturally and genuinely increase your use of this skill, observe what it does for the communication process.

Poor Substitutes For Responding with Understanding

Responding with accurate understanding looks simple on paper, but this simplicity is deceptive—as you're probably finding out if you've tried to practice this skill. Often enough, when accurate

155

understanding is called for in communication, you'll find tendencies in yourself to give other, less helpful responses. Let's take a look at some of these kinds of response.

Clichés. Usually, responding to another person with a cliché is worse than not responding at all. Clichés are phony, and they put distance between communicators. For example, suppose that a member of your group says, a little nervously:

> "We haven't talked about sexual feelings here at all. If I were being honest with myself, I think I'd have to admit that in some ways I'm a bit sexually immature. I've thought about it a lot during the group meetings, but I haven't said anything. And I've been very cautious with the women here. I think that my guard has been up with the women, and I'll bet some of you have noticed it."

Now consider the following responses.

Bert: "Many people struggle with sexual problems throughout their lives, John."

This response may sound silly, but don't laugh too soon. Clichés like this come up in groups, and they seem to abound in everyday life. Here is another such response.

Denise: "I think I understand what you're saying, John."

This is a commonly heard cliché in groups. Sometimes it's simply a lie, because the person who says it doesn't understand at all but feels that he or she should say *something.* In a sense, clichés are disrespectful, for they are ways of *not* responding to the speaker. They don't move the conversation ahead at all. How would you respond to John's statement?

_____

_____

One way of encouraging mutuality in the group is to respond with accurate understanding *plus* a willingness to join with the speaker in examining an issue. For instance:

"John, even though it's rather embarrassing and anxiety-provoking, you think that it might be a good idea to take a look at the issue of sexuality in the group directly. Actually, none of us has said much about sex. Maybe it's an issue we could all examine, at least insofar as it influences us here in the group."

To repeat: clichés don't promote mutuality; instead they actually place some distance between the speaker and the person responding.

Parroting. Parroting is a mechanical restatement of what another person says. Let's take an example.

Alice: "Dave, I just dumped my anger on you, but now that I've had a moment to think, it doesn't seem fair. I'm actually angry at myself for running away from confrontations that are quite legitimate and useful. Attacking the confronter is one way I run."

Dave: "You don't think it was fair to dump your anger on me. You're really angry with yourself for running away from legitimate and useful confrontations. Attacking me was a form of running."

Dave's response could have been given by a tape recorder. It lacks humanity; it lacks a sense of mutuality. Responding with understanding means genuinely getting inside the other person to the degree that you can. It means that you not only listen but also make the viewpoint of the other person your own. A good communicator is not listening just to words; he or she is always actively listening for the core of what the other person is saying. The good response gets at this core and isn't merely a repetition or paraphrase of what the speaker says. How would you respond to Alice?

_____

_____

Inadequate responses. Some people think that they're good listeners because they pay attention to the speaker and give little cues that they're paying attention, such as nodding their heads or saying "uh-huh." However, if the speaker really discloses something about himself or herself, such minimal responses are often inadequate. Let's take an example.

> *Sue:* "Ken, I don't think that you and I are getting anywhere. We go out a lot, and generally we have a good time. But maybe there's something wrong with me. So often I feel that I'm with my brother. You know, I like my brother, but I want to feel that you're more than a brother. I don't know exactly how to put it. And I don't want to blame you or just dump this in your lap."

> *Ken:* "Uh-huh."

The trouble with an inadequate response is that it can make the speaker feel as if he or she hadn't said anything worth responding to. In this case, Ken might actually be very interested. He might be feeling close to Sue, or he might be all torn up inside thinking about how ambiguously he feels about her. But his "uh-huh" simply isn't enough. Conversations are usually effective to the degree that they are *dialogues*. If Ken lets Sue talk on and on without responding adequately, when she stops he might start on a long monologue of his own. Obviously Ken and Sue need to dialogue; they need to reveal themselves and to respond to each other in shorter "chunks" of conversation if they're going to face and solve the issue that is separating them. If you were Ken, how would you respond to what Sue said?

Inaccurate understanding. Understanding communicated to others is usually helpful only if it is *accurate*. If you communicate inaccurate understanding, the other person will probably let you know in one of many ways. He or she may stop dead, not knowing how to handle your inaccurate response, or else fumble around, go off on a new tangent, tell you "That's not exactly what I meant," or use some other means to let you know that your response wasn't really helpful. Here is an example.

> *Tom:* "Maureen, I really didn't want to give you the impression that you're butting into my life. You call me on it when I get moody and withdraw in the group. I find it hard to admit

to myself that I'm still wrestling with childish ways of expressing emotion, and often enough I don't take confrontations very well. But I still want to be in the group and face these issues."

Maureen: "You see me as telling you to get rid of your childish ways. And you resent me for it."

What's wrong with Maureen's response? Well, it isn't accurate; that's not what Tom said. In this case it's more than likely that Maureen is being overly sensitive and letting her own feelings interfere with her response to Tom. If what Tom is saying is disturbing her, it would be more effective if she dealt with her own emotions instead of letting them get in the way of responding to Tom. For instance, she might have said:

"Tom, I'm getting upset. I think I hear you saying 'Stop, stay out of my life, you're hurting me.' But I'm so emotional right now, I doubt that I'm getting the right message."

This response gives Maureen and Tom a chance to clear the air before they get bogged down in mutual misunderstanding. How would you respond to what Tom had to say?

_____

_____

Pretended understanding. Sometimes it will be difficult to understand what another person is trying to communicate, even though you're giving him or her your complete attention. If the person is confused, distracted, or in a highly emotional state, the clarity of what he or she is trying to say may suffer. There may also be times when you yourself become distracted, get lost in your own thoughts, or become confused—times when you'll fail to follow what someone is saying. It's against the spirit of understanding to pretend at such times that you do understand. It's more genuine to admit that you're lost and to work on getting back on track: "I've gotten distracted and lost you. Could you repeat what you just said?" If you're confused, admit your confusion: "I'm sorry. I don't think I followed what you just said. Could you go through it once more?" Such statements are signs of your respect, since by them you say that it's important to you to stay with the speaker and to genuinely understand him or her.

Admitting that you're distracted or lost is much preferable to muttering such clichés as "uh-huh" or "ummmmm" or "I think I understand."

If you're not quite sure that you understand what the other person is trying to express, be *tentative* in your communication of understanding; that is, let your response show that you aren't absolutely sure. Your being tentative gives the other person room to move, so that he or she will feel free to correct you or give you a clearer picture of what he or she really means. Let's take an example.

> *Peter:* "If I understand you, you think that I find it difficult to trust you because your emotions change so rapidly. Is it something like that?"

> *Debbie:* "That's partly it. I don't even trust *myself* when my mood changes so quickly. And I guess that I don't expect to be trusted."

Peter's response enables Debbie to clarify what she means.

A good technique to use instead of pretended understanding is perception checking. Perception checking means asking the person you're talking with whether the understanding you're communicating is accurate or not. Perception checking keeps you on track in your dialogue with another person. There are a number of ways of going about perception checking, as shown in the following examples. The italicized words are the ones that indicate that you're not sure, that your response is tentative.

- "It *sounds like* you're angry both with me and with yourself. *I could be wrong, but that's how I feel.*"
- "It *seems to me* that you'd like a little time to think about what I just said. *Am I reading you correctly?*"
- "*I'm not absolutely sure, but I think* that you're saying to Joan that you like her but that she also does things that annoy you. *Does that hit the mark?*"
- "*I have the feeling* that you're angry with me but that you're holding it back. *I'd like to check this out with you.*"

Perception checking helps prevent misunderstandings and thus keeps the dialogue flowing freely. It helps get rid of any tendency to be judgmental in the communication of understanding and is therefore a sign of respect. However, if you overuse perception check-

160

ing, your conversation will become watered down and weak. Don't use it merely because you're afraid to talk to the other person directly. If that's the case, it's much better to deal directly with your fear.

Ignoring of what a person says. This is a type of inadequate response. Some people listen to another person and then simply either say nothing or change the subject.

> *Todd:* "Something funny is happening to me. Outside the group I'm much more open than I used to be. I reveal myself much more to my friends. They've even told me so. But I'm not revealing very much about myself in this group at all. It seems that I'm *learning* in here and *doing* outside."

> *Steve* (after a short pause): "I'd like to get some feedback on my fight with Fred last week."

A person who fails to receive some kind of constructive response to what he or she says is less likely to risk anything in the group later on. It's better to let a person know that you don't know *how* to respond than simply to refrain from responding at all.

Long-windedness. Conversations, either one-to-one or in groups, are usually more engaging if there is a great deal of dialogue. Long-winded monologues followed by equally long-winded replies are deadly. They rob the group of its vitality. Part of the art of responding with understanding is making your responses relatively short. A "lean" response is usually much more effective than a "fat" one. Besides, it usually takes more words to say nothing than to say something. In one group, two members who were having trouble communicating with each other were trying to deal with the issues that separated them. When they began talking, they felt awkward and didn't know what to say. Finally, they arrived at the following exchange.

> *Tony:* "I can say that I like you, but with the qualification 'as far as I know you.' I don't know you too well, but I like what I see."

> *Ron:* "I like you without qualification."

Ron's remark was short, but it had great impact.

In trying to be accurate, the beginner may tend to be long-winded, sometimes speaking longer than the person he or she is responding to. It's best not to try to respond with accurate understanding too quickly. Giving yourself a chance to think about what you want to say will help you make your responses lean instead of fat. In the following example, the person responding jumps in too quickly, realizes that his first few sentences are wide of the mark, and then keeps on talking in hopes of eventually being accurate. Quantity takes the place of quality.

> *Cindy:* "I've never been very spontaneous in social situations, and this group is no different. I'm shy; I stay out of things until I see where I fit. In conversations I wait so long to get in that, by the time I have something to say, the conversation passes me by. While the others are talking, I'm just too preoccupied with myself—how I'll be accepted, how to say the right thing—all that sort of stuff."

> *Paul* (jumping in right away): "You're really shy, and that cuts down on your being spontaneous. It shows up in groups like this one. You're listening all right. You know what people are saying. But then you begin to ask yourself 'What should I say? I shouldn't sit here dumb like this.' But, by the time you think of what to say, it's just too late. No, it's even worse than that. You get lost in yourself, and then it's twice as hard to get back into the conversation, because by the time you tune back in you're not sure where the others are. Your shyness backfires on you in more than one way. And then you sit there even more out of it than before."

Paul's response may well be accurate, but it's just too much. It places emphasis on his own attempt to understand rather than on what is bothering Cindy. Cindy may well be smothered by all that Paul has to say and may not really feel encouraged to move on. How would you respond to Cindy with accurate but lean understanding?

If you keep asking yourself "What is the *core* of what this person is saying?" you can make your responses short, lean, concrete, *and* accurate.

162

Questions. Questions are usually poor substitutes for responding with understanding. Consider the following example.

> *Harry:* "I think that I'm a very affectionate person. I like to be close to people, both male and female. I like physical affection, and I like psychological closeness. But I'm very fearful of expressing affection in almost any way. I'm afraid that people will reject me. I'm afraid that they will see me as soft or unmanly. I know that's a lot of crap, but it still inhibits me from showing affection. It all gets locked up inside of me. If you told the people who know me outside the group that I'm an affectionate person, they'd probably laugh, because they never see it."

> *Frank:* "What do you do when you feel affectionate and don't express it?"

Frank's question is an extremely poor response to Harry. Harry has just exposed a vulnerable and frightening side of himself, and he deserves a better response. Instead of understanding, he gets a question. Asking a question at this time is almost the same as saying "Harry, you haven't really said anything about yourself, so I'm going to ask a question to get the *real* information." It is a form of disrespect or disregard for Harry's self-disclosure. It's even poor counseling practice to ask a lot of questions, and they're that much more uncalled for in mutual conversations between friends. People who don't know how to respond with understanding usually ask a lot of questions instead. In the example just given, Frank's question is really socially *un*intelligent. What's worse is the fact that, once you begin to ask questions, it's difficult to stop. Questions beget questions, and then all mutuality is lost. The point is not that you should never ask a question but that it's much better to avoid inappropriate ones. Frank's question is inappropriate because accurate understanding is what Harry's self-disclosure calls for. How would you respond to Harry?

_____

_____

Asking a lot of questions turns the group session into a game of psychology or of counseling. Overemphasis on questions is one of the most destructive forms of group participation.

Interpretations. Another poor substitution for understanding is giving an interpretation of what the speaker says. Consider the following example.

> *Shannon:* "I'm never really satisfied with myself. For instance, I'd like to be very popular with the other students. I *do* have good friends, but I'm not really what you'd call popular. I'd also like to be good at some sport like tennis, but I'm just average. I could give a lot of other examples. I'm just *too* average in almost everything, and I don't like it."

> *Bruce:* "I'll bet that your real problem is that you haven't received enough attention from your parents. If you had, you'd be more satisfied with yourself."

Bruce's response isn't useful. It doesn't help him get closer to Shannon. Interpretations like this one are part of the psychologist game, which is a poor game to play in interpersonal relationships. Playing psychologist destroys mutuality; it creates distance. It makes one person the helper and the other the client. How would you respond to Shannon?

_____

_____

Judgments. I've already mentioned our instinctive tendency to judge and evaluate what the other person says instead of first merely trying to understand the person and communicate this understanding. Letting our tendency to be judgmental creep into our responses does little to develop mutuality. The judge is one up, and the person judged is one down. Let's take an example.

> *Kevin:* "When I play games, I play them hard. The object of a game is to win, and that's what I try to do. I do the same in relationships. If I want to relate to someone, I push hard in the relationship. I hate boring, lifeless relationships, so I try to stir things up between me and others. I think a lot of people like my drive. I feel the same way in this group. We've got goals, and I want to see us get to them."

> *Rick:* "Kevin, you're too competitive. You define all of life as competition. You wear people out with your competing."

By this response, Rick sets himself up for an argument with Kevin. It may well be that Kevin's competitiveness is self-defeating in interpersonal situations, but merely accusing him of being overly competitive is an ineffective way of facing the issue. As we will see later, if you want to challenge or confront someone, you have the right to do so to the degree that you understand him or her first. Understanding comes before confrontation and paves the way for it. The relevant question is not whether Rick's evaluation is right or wrong but (1) whether he has spent enough time trying to understand Kevin and (2) whether the way he is confronting him is likely to be effective. How would you respond to Kevin?

Advice giving. Advice giving is another common substitute for the communication of understanding. Most of us fall into the trap of giving others advice, even though they don't ask for it and might even resent it. Giving advice doesn't promote mutuality. The advice giver is one up, while the other person is one down. Here is an example.

> *Lynne:* "Things have been looking up for me. I've stopped being aggressive in most of my relationships. For instance, my boy friend is a lot more relaxed when he's with me, because I'm not pushing him to do things. I'm really glad I found out the difference between being assertive and being aggressive. I thought that giving up being aggressive would mean that I'd become a mouse or someone everyone else would walk on. But that hasn't been true at all, either in the group or outside. I'm getting my legitimate needs met, but I'm not walking over other people."

> *Michele:* "Don't tone down too much. Remember that women get walked on if they give in even a little. You should be more confronting in this group."

Even if Michele is convinced that Lynne has moved from being aggressive to being nonassertive at times, giving Lynne advice is parental behavior that isn't likely to help build their relationship. Michele's advice may even be sound, but that's not the issue here. The issue is that she didn't take time to communicate any understanding

165

of Lynne's hope and enthusiasm. How would you respond with understanding to Lynne?

—————————————————————————————————

—————————————————————————————————

If you're an advice giver, you might ask yourself: Do people take my advice? Does giving advice help me to promote mutual friendships?

Patronizing responses. Another poor substitute for communicating understanding is communicating a patronizing or condescending attitude. The speaker is seen as a poor child, while the responder is the understanding parent. Again, this is a one-up/one-down situation, as illustrated in the following example.

> *Clare:* "I really let my fears get the best of me when I'm not sure whether the people I'm with accept me completely. It happens here in the group. I'm afraid of you, John, and, since I see you as strong, I'm very quiet around you. Marsha, I'm afraid of your wit and your tongue, and so I see myself playing up to you like a little girl so that you won't hurt me. The worst part of this is that I don't really see myself making much progress."

> *John:* "Oh, Clare, I know that you've got your fears, but you're doing just fine. Just tell us when you're afraid. And give yourself time!"

Clare has just said that she dislikes it when she acts like a fearful little girl, and then John responds by treating her exactly like a fearful little girl. His response is disrespectful; it's patronizing and condescending. He doesn't even recognize that Clare is quite dissatisfied with herself. John's response is a sexist male response. How would you respond to Clare?

—————————————————————————————————

—————————————————————————————————

Patronizing responses need not be as obvious as John's. Even hints of patronizing and condescending attitudes can destroy mutuality.

Defensiveness. When you feel that you're being attacked or confronted unfairly by the remark of another person, your instinct may well be to become defensive or to fight back. Such self-protective responses are certainly understandable, but they usually do little to promote effective communication and mutuality or to build relationships. Let's take an example.

> *Carl:* "I get the feeling that some of you are talking to one another outside the group. I see knowing glances and I hear comments passed that sound like 'in jokes.' I don't see any of you between meetings, and every time I show up here I feel in many ways like a stranger. I've got better things to do than to try to get into an 'in' group. Sandy, you've talked about your telephone conversations with Matt. I'm not sure how much of that goes on."

If the group members see Carl's comments primarily as an attack, they might react in a number of ways.

> *Sandy:* "I resent the implications of what you're saying. I'm honest and straightforward here. I'm not hiding anything or playing games. And I have a perfect right to a private life outside the group. I didn't give that up by coming here."

Sandy responds by vigorously defending herself. You may say that she has a right to defend herself, but you might also ask whether her self-defense is the most constructive or creative response. Defending herself merely accentuates the distance introduced by Carl's remarks. If Sandy first communicated to Carl an accurate understanding of what he's feeling, Carl might be more likely to examine his own suspiciousness.

> *Ned:* "You're one of the most suspicious people I've ever met. Everyone here has to handle you with kid gloves, because you always think that people are out to get you one way or another. Get off it! Grow up and join the human race!"

Whereas Sandy responded with self-defense, Ned takes the other possibility—he vigorously attacks Carl. It goes without saying that the distance created by Carl's remarks is made even larger by Ned. You might say that Ned has a right to get angry and to dump his hostility on Carl. Be that as it may, his response is not particularly constructive or creative. Accusing, punishing, and belittling Carl doesn't solve the problem; it only complicates it. When you feel attacked, it will be quite difficult for you to respond first with some kind of understanding of the attacker's feelings and of why he or she feels that way. You'll find such a response impossible unless responding with understanding has become second nature to you. I'm not saying that you should surrender your right to get angry or your right to challenge others. You don't have to surrender any of your rights. However, the communication of accurate understanding is the best means of creating an *atmosphere* in which people can work out their differences. Let's take a look at another possible response to Carl.

> *Patricia:* "Oh, come on, Carl. You know that's not the case. I like you. We've had some excellent interactions here. I know you don't think that I'm doing anything behind your back. I mean, don't you think that we get along well?"

Patricia is trying to *manipulate* Carl, to sweet-talk him, to win him over to her "reasonable" point of view. However, such manipulation is disrespectful, for it ignores both Carl's feelings and the issue he's raising. Patricia gives the impression of trying to overcome interpersonal distance, but her tactics are questionable. If Carl lets himself be sweet-talked, it's quite likely that he'll merely swallow his anger and his suspicion for a while only to have them erupt, perhaps even more violently, later on. If your goal is to understand Carl and how he feels, how would you respond to him?

_____

_____

The Manner of Communicating Understanding

Here are a few hints that will help make your communication of understanding as effective as possible.

1. *Use plain and understandable language.* Your communication of understanding will probably be better received if your language is in tune with the language of the person you're responding to. If your language is very formal and stiff, while the language of the other person is very casual and slangy, your language might well get

in the way of your communication of understanding. To illustrate, let's take a relatively silly example.

> *Group Member A:* "Man, I'm spaced out right now. I'm not sure you cats even know where I'm coming from. I'm about to blow my cool. This group stuff is a psychological rip-off."

> *Group Member B:* "You feel frustrated and angry because our perceptions of you are not really empathic."

These two people are worlds apart in terms of language. However, you can be yourself and yet accomodate yourself in some degree to the other person's language, as shown by the following response.

> *Group Member C:* "You feel that nobody here is picking you up, and that's lousy."

The point to remember is that a style of language is itself a form of communication. Two people who want to communicate effectively would do well to find some common ground in terms of language.

2. *Make your tone and your manner of speaking communicate also.* I've already discussed how your voice quality sends messages just as well as your words. If someone speaks excitedly about her involvement with her training group, telling you how great it feels to be establishing some solid relationships, and if you indicate in a flat, dull, lifeless voice how glad you are to share in her joy, your response will be practically useless—even if it's accurate. She'll hear the quality of your voice and not the message of your words. This doesn't mean that you should mimic the speaker's enthusiasm (or whatever the emotion being expressed happens to be), but there should be some proportion or congruence between your tone and manner of responding and those of the person you're responding to. Responding well means *being with* the other, and that includes the quality of your voice.

3. *Give yourself time to reflect.* Good communicators don't immediately leap in as soon as the person they're talking with pauses. They give themselves a few moments to think, often because they're trying to digest what the other person has said and because they think it's important to get at the core of what is being said. If you leap in too soon, chances are that you won't be accurate or that you'll be long-

winded. I'm not implying that your conversation should be filled with uncomfortable pauses; you can be reflective and still maintain your spontaneity.

4. *Respond frequently.* Don't save up your understanding and deliver it all at once. You probably won't be accurate, and you'll certainly be long-winded. People who respond to each other relatively frequently in a conversation tend to keep the conversation mutual and spontaneous. Although you shouldn't go to the extreme of continually interrupting, it's too easy to let the other person go on and on. Then, when he or she eventually stops, you may well remain silent—not because you don't want to respond but because you don't know where to begin.

5. *Both give understanding and expect it from others.* A high-level communicator responds frequently and with understanding. However, he or she also expects to receive understanding from others. If others fail to give you understanding, you have a right to ask for it:

> "I've been talking for a while now. I'd like to get some reaction to what I've been saying. I'm not sure how I'm coming across."

Part of assertiveness is asking for what you need in order to communicate well. Inviting response is one way of avoiding mutually dissatisfying monologues.

6. *Respond to feelings and/or to content.* Don't feel that you must *always* respond to both feelings and content. In the examples so far, good responses have stressed both, and you've also been asked to respond in this way in the exercises. However, at any given time in a conversation, one or the other might be more important. Here is an example.

> *Barbara:* "What's going on here? We're all good at the skills we've been learning, but we're still not going anywhere. Our relationships seem lifeless. Most of you respond well to me, but there's still a lot of distance between us. I don't feel any movement in the group, damn it, and I know that it's just as much my fault as anyone else's."

> *Mike:* "You're really fed up."

> *Barbara:* "You're damn right I'm fed up. And I'm hoping that I'm not the only one who's fed up. Let's see whether other people feel the same distance as I do."

Mike responds only to Barbara's feelings, but his response hits the mark closely enough to encourage her to move along. It's also possible to respond occasionally just to content:

> *Tim:* "Norm, when Jean is silent, it bothers you, and you let her know pretty quickly. But, when I'm silent, you let me alone. I think I could name other times when you seem to deal with people quite differently."

> *Norm:* "I must seem pretty inconsistent to you."

> *Tim:* "Yes, you do. But I keep wondering why it is that you treat people so differently. I don't know enough about what you feel about Jean that enables you to call her on her silence. And I guess I don't know how you feel about me."

Norm listens, and he responds nondefensively to *content*— that is, to Tim's experiencing of Norm—rather than to feelings. But Norm seems to hit the mark, because Tim goes on to become more specific.

If a group member shows early in the life of the group that he or she is easily threatened by a discussion of his or her feelings, you may want to emphasize content when you respond to that person in the beginning and then gradually deal with feelings more and more. One relatively unthreatening way to encourage another to deal with feelings is to indicate what you might feel in similar circumstances.

> *Carol:* "A lot of responses I get here seem to assume that I'm a little kid. And I'm in my mid-30s! I get encouraged to talk. I get gentle reprimands for not keeping up with the contract."

> *You:* "I think that, if I were being treated that way, I'd be pretty angry."

Of course, in this case such a response could backfire if Carol saw it as just another instance of being treated like a child.

The "trust package" and the "trust spiral." It's always to some degree a risk both to share yourself and to provide understanding for others in a group or in any relationship. If you disclose yourself, you make yourself vulnerable. In a sense, you deliver yourself over to the hands of others. And, if you try to respond to others with understanding, not only do you have to work at attending, listening, and responding, but you risk failing to understand. You brave the closeness or intimacy that comes from entering another's world.

It's meaningless to ask which is more important—self-disclosure or understanding. The two together make the "trust package." You'll find that your self-disclosure, if it is appropriate, will have a way of stimulating others to reveal themselves. Just as your trust will encourage them to trust you, so their trust will enable you to trust them even further. Then, if you provide accurate understanding and receive it from others, you and the other group members will launch yourselves on the "trust spiral."

Exercise 31
Identifying Poor Substitutes for
Understanding

Before beginning this exercise, let's briefly review what some of the common substitutes for responding with understanding are:

- clichés
- parroting
- inadequate responses ("uh-huh," nods, vague statements, unrelated remarks)
- inaccurate understanding
- pretended understanding
- ignoring of what a person says (saying nothing, changing the subject)
- long-windedness
- inappropriate and repeated questions
- interpretations, playing psychologist
- judgments
- advice giving
- patronizing or condescending responses
- defensiveness (defending, attacking, manipulating)

This list probably doesn't exhaust the number of poor substitutes for communicating understanding; perhaps you can think of others. Some of the errors just listed are demonstrated in the exercise. You're asked to identify them.

Directions

In this exercise, a number of statements by people trying to communicate what they think and feel are given together with a number of possible responses.

a. If the response is good—that is, if it's a communication of accurate understanding—give it a plus sign (+). However, if it's an inadequate or poor response, give it a minus sign (−).

b. If for any reason you give the response a minus, indicate briefly *why* it's a poor or inadequate response. A response may be poor for more than one reason. Make your reasons as specific as possible.

c. Give your own response to each statement.

Example

"I have high expectations of this group. I think that we've developed a pretty good level of trust among ourselves, and frankly I'd like to start taking bigger risks. I'd like to be able to talk about myself more freely. The longer I'm here, the more I desire to learn as much as possible about myself and my interpersonal style. I'd like you to help me take more risks, and I'd like to do the same for you."

a. (−) "Hey, I wish you wouldn't speak for me. I'm not too sure that my expectations are the same as yours. I think you might be too idealistic for me."

    Reason: defensive, judgmental, accusatory

b. (+) "Your enthusiasm is really growing. You think that we could all help one another move deeper."

    Reason: (None, because it's a plus.)

c. (−) "Do you think we're really ready to do that sort of thing?"

    Reason: inappropriate question, vague, defensive

d. (−) "Now, Leslie, you've been a very good member— very eager. I appreciate your eagerness very much, but don't let your enthusiasm run away with you. Make haste slowly."

    Reason: condescending, parental, advice-giving

173

e. (+) "Your enthusiasm's contagious, Leslie; at least it is for me. There's something of the coward in me, but maybe I'm ready for a little more risk-taking myself."

Reason: (None, because it's a plus.)

Now complete the following.

1. "I know I've been pretty silent. I didn't feel right barging in on Paul and Marie's conversation, so I waited until I thought they were finished. I keep thinking that people will get angry if I interrupt. Maybe it's the wrong way to be, but I don't interrupt people outside the group, and it's hard for me to think that it's OK here."

  a. ( ) "It seems that you're afraid of being rejected if you interrupt. And rejection really hurts you, because you don't see yourself as a worthwhile person."

  Reason: _____

  b. ( ) "I think you're being very unfair; you're not giving Paul and Marie much credit."

  Reason: _____

  c. ( ) "Peter, you know our contract. What you call 'barging in' is just moving into a conversation. Since you're timid, I think you should make extra efforts to push your way in."

  Reason: _____

  d. ( ) "Moving into an ongoing conversation just doesn't seem right to you yet. So it's really hard for you to move in."

  Reason: _____

  Your response: _____

  _____

2. "I think my skills have improved a lot in the last two or three weeks. I'm able to express my feelings much more openly and honestly. And feeling more confident has helped me to become less defensive when I talk to others."

a. ( ) "But you feel half finished, because you haven't been able to lick your defensiveness completely."

*Reason:*_____

b. ( ) "Yeah, I can see that."

*Reason:*_____

c. ( ) "How have you managed to become less defensive?"

*Reason:*_____

d. ( ) "I know that your skills are improving. I've seen that every week. But I can't say that I see you as less defensive."

*Reason:*_____

*Your response:*_____

_____

3. "I see myself as a very independent guy. My independence, I think, tends to rub some people the wrong way. Some people think that they can't affect me in any way. I'm not saying that being this independent is right or wrong, but it is the way I am. I think it's only fair to let you people here know. Don't look for something I don't usually give."

a. ( ) "I think I understand. I used to think that I was more 'together' than anyone else I knew. Other people didn't matter to me."

*Reason:*_____

b. ( ) "I think I understand independence. I'll bet that somewhere along the line you let someone get close, and you got stung. But I think you can trust us enough to let yourself go a little."

*Reason:*_____

c. ( ) "Right now you feel pretty strongly about being independent. I'm not sure what else you feel about it, except that you think it might cause some trouble here."

*Reason:*_____

d. ( ) "I don't see you that way at all. I experience you as open and 'with' us. I like being in the group with you."

Reason: _____

e. ( ) "Well, isn't that nice! You just want to be left alone—maybe because you're afraid of being dealt with here."

Reason: _____

Your response: _____

_____

4. "Janet, I'm having a hell of a time understanding you. It could be that I'm putting up some kind of wall between us. I know I've been defensive with you a couple of times. I get the idea that you avoid talking with me, but I'm reluctant to say that it's your problem. I guess I don't know whether it's you or me or both of us."

a. ( ) "It's good to bring things up like this. I think I know what you're driving at."

Reason: _____

b. ( ) "I like the way you can look at yourself and admit your own faults and your part of what goes wrong in a relationship."

Reason: _____

c. ( ) "You're confused and angry right now, because you don't know how to deal with my being distant."

Reason: _____

d. ( ) "I think it's rather unfair, putting it on me like this. I try not to run away from facing any relationship. I think my behavior in the group has shown that."

Reason: _____

Your response: _____

_____

5. "I don't believe what's happening to me. I came in here six weeks ago expecting to rap or relate or something like that, and now I find my whole interpersonal style chang-

ing. The most satisfying thing is the change that's taking place between me and my husband. We listen to each other. We respond to each other. I never expected so much out of a course."

a. ( ) "I'd like to take a look at what's happening between you and me, Ann. I feel that there is still some unresolved hostility between us that should be taken care of before going into what's happening to you outside."

Reason: _____

b. ( ) "I'm not sure that I understand. How exactly has your marriage changed because of the group?"

Reason: _____

c. ( ) "It sounds like you could hardly be more pleased with the group—especially with the fact that it could affect your closest relationship so positively!"

Reason: _____

d. ( ) "You came in here expecting to rap or relate, but you find your whole interpersonal style changing. The most satisfying part is with your husband. You listen and understand. You're getting a lot out of this course."

Reason: _____

Your response: _____

_____

6. "I'm not a very understanding person, and I can't change overnight. I don't see why we have to stress understanding so much. It's phony. People don't talk like that to one another. I don't see what's wrong with free-flowing conversation. It's getting so stilted here that I wonder whether the course has any application to relationships outside. I bet you people don't 'respond with accurate understanding' outside any more than I do. Why don't we admit it and get on with relating?"

a. ( ) "I think you've hit the nail on the head. If our attempts at understanding are so artificial, they can't do much good for us in our day-to-day relationships."

Reason: _____

b. ( ) "Please, Mike. I wish you wouldn't push your own views so much. I get a lot out of these practice sessions. You don't like the idea of accurate understanding because you're not good at it. Don't dump your problems on the rest of us."

Reason: _____

c. ( ) "This group isn't real life; practicing accurate understanding and stressing it so much doesn't seem realistic, since, if we admit it, we don't use it very much outside. So you're frustrated. You'd like to see us get down to some more realistic issues. You'd feel more comfortable with less practicing and fewer rules. It's not that you're saying that the group has nothing to offer. Just relating is OK, and you'd like to do more of it."

Reason: _____

d. ( ) "I like what we're doing. I find that I'm communicating understanding more and more at home and that it works for me."

Reason: _____

Your response: _____

_____

7. "I'd like to start by saying that it's a brand-new experience for me to receive criticism and confrontation in such a way that I don't have to get defensive. I'm used to defending myself, but most of you challenge me in ways I really like."

a. ( ) "Good. [Pause.] Aren't there some unresolved issues from last week?"

Reason: _____

b. ( ) "Bill, you said 'most of you.' Who in here challenges you in ways that make you defensive?"

Reason: _____

c. ( ) "You feel that the give-and-take in here has been pretty constructive—and that's a good feeling."

Reason: _____

d. ( ) "I think that I find constructive confrontation as rewarding as you do, Bill. But I don't get much of it from you."

   Reason: _____

   Your response: _____

   _____

Exercises in Trust

Exercise 32
Physical Trust: The Trust Walk

This exercise will let you see what it's like to entrust your physical safety to another person. You'll get a deeper feeling for the process of entrusting yourself.

Directions

a. Choose a partner.
b. Either you or your partner closes his or her eyes and keeps them closed during the exercise.
c. The "blind" person entrusts himself or herself to the other physically, allowing himself or herself to be led around the room or down corridors.
d. The "seeing" person can have the "blind" partner do a variety of things—walk around objects, run down corridors, climb on chairs, and so on. The "seeing" partner, however, in no way allows the "blind" partner to stumble, run into things, fall, or hurt himself or herself in any way.
e. Allow five minutes or so for the "seeing" partner to lead the "blind" partner around.
f. Now the two of you trade roles and continue the exercise for another five minutes or so.

Discuss with your partner how it felt to entrust yourself physically to him or her. Did you move easily with your partner,

179

or were you very hesitant? How did you handle your fears? Did anything happen that weakened your trust?

Discuss how you feel about entrusting yourself to the group in other ways—for instance, through self-disclosure. Is it easy for you to entrust yourself to others psychologically? What's easy about it? What's hard about it? What do you have to see in another person to allow yourself to trust him or her? How do you feel about entrusting yourself to this group of people?

Exercise 33
Physical Trust: Being Passed Around

This exercise will help you see what it's like to entrust your physical safety to a group of people and will enable you to get a deeper feeling for the process of entrusting yourself to a group of people.

Directions

a. Five or six members of the group form a tight circle, all members standing.
b. Another member of the group stands in the middle, encircled by fellow group members.
c. The member in the middle "free falls" toward the members encircling him or her.
d. The members in the circle don't allow the person to fall to the floor or to hurt himself or herself in any way. The members hold the falling member up and pass him or her around from member to member (or to two members if one member cannot assume the total burden). The members cooperate in passing the falling or limp member around the group.
e. Each person takes his or her turn being passed around.

After each member has had a turn, the members sit down and discuss the experience. How did you feel falling? How did you feel being passed from member to member? How easy or difficult was it for you to entrust yourself physically to the group? How did you feel about being touched by the group members as they passed you from person to person?

For you, is there any relationship between this kind of entrusting and entrusting yourself psychologically to this group of people? What difficulties do you have in entrusting yourself psychologically to them? What rewards do you find in entrusting yourself to them?

Genuineness and Respect as Communication Skills

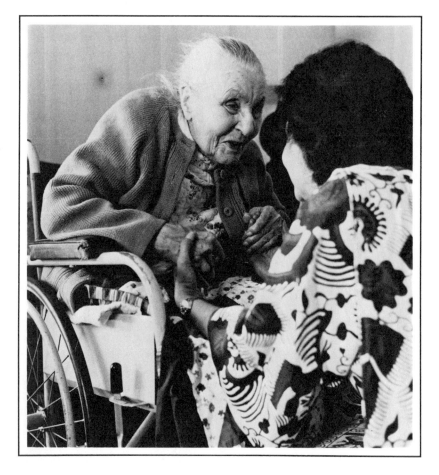

Most people think of genuineness and respect as values or moral qualities or attitudes rather than as skills. However, even as values or moral qualities, genuineness and respect make sense only to the degree that they are translated into interpersonal behaviors. This chapter discusses how the group experience can help you to see (a) to what degree genuineness and respect are interpersonal values for you and (b) what the behaviors are that constitute these values.

Genuineness: A Behavioral Approach

On the supposition that you want to be genuine in your interpersonal relationships, how do you go about expressing genuineness? We all have some feeling for what genuineness means, but in this section I'll try to make this concept much more concrete by talking about the *behaviors* through which genuineness is expressed.

1. *Freedom from roles.* A genuine communicator doesn't hide behind roles. To be sure, roles are important in human life. For instance, they help us set up expectations in social situations. You expect a waiter or waitress to serve you food, not to fix your car or solve your problems. And, although you might be friendly with a waiter or waitress, you don't behave toward him or her as you would toward a good friend. However, more intimate interpersonal communication seems to be at its best when it is role-free. For instance, if I'm a teacher and I still act like a teacher when I come home to my wife, then I'm not being genuine. I'm hiding behind my role as a teacher. Being role-free in closer relationships is one way of being genuine. How does a role-free person act toward others? He or she:

- expresses directly whatever is going on inside him or her if it's appropriate to do so;

- communicates directly without distorting his or her messages;
- listens to others without distorting their messages;
- reveals his or her true motivation in the process of communicating whatever he or she has to say;
- doesn't plan out what is to be said to the other person;
- doesn't use planned strategies to deal with people;
- responds immediately to the other person's need instead of waiting for the "right" time or giving himself or herself enough time to come up with the "right" response;
- lets himself or herself appear vulnerable or weak if that is the case at any particular time;
- communicates in the here and now;
- strives for interdependence in relationships rather than dependence or antagonism;
- learns how to enjoy closeness and intimacy;
- is concrete in his or her communications;
- is willing to commit himself or herself to others.

Such a person isn't just a free spirit who inflicts himself or herself on others. "Being role-free" doesn't mean the same thing as "letting it all hang out." It does mean not using disguises to fool people and not using games to manipulate them.

Let's put some of this theory into an example. If you, as a group member, fear the group trainer or instructor but refuse to talk about this fear with either the instructor or your fellow group members, if you're afraid to initiate anything without the instructor's OK and you see the instructor mainly as an authority figure, if you're either dependent on the instructor or antagonistic toward him or her, then you are violating some of the principles outlined above. You're locked into the role of pupil or trainee. The trainer or instructor does have a legitimate role insofar as he or she possesses the skills you want to learn and knows the methodology by which you can be helped to learn them. You, too, have a legitimate role insofar as you're trying to learn these skills. However, if the instructor uses his or her role to keep from getting involved with the students, and if the students use their role as an excuse for being passive, dependent, or uncooperative, then instructor and students are both locked into self-defeating roles. Neither is role-free in the sense described above.

2. *Spontaneity.* Spontaneity is a sign of genuineness. An effective communicator, while being tactful and respectful of the rights of others, is not constantly weighing what to say. In revealing himself or

184

herself and in responding to and challenging others, an effective communicator is assertive without thereby stepping on anyone's rights. If you're overly controlled outside the group, you can use the group to learn that it's all right to be active, spontaneous, free, and assertive. In my experience with human-relations-training groups, lack of assertiveness in communication is a very common problem.

A spontaneous person is free without being impulsive. As we've already seen, a spontaneous person doesn't live haphazardly, doing whatever comes to mind, but rather lives a disciplined life that enables him or her to move freely. A good communicator does learn some basic rules, but, more importantly, he or she learns some basic *skills*. Like the disciplined basketball player I talked about earlier, a high-level communicator has many skills and responses at hand. Thus he or she has the ability to be spontaneous—that is, to call on any communication resource as the need arises.

3. *Nondefensiveness*. Good communicators can be genuine in handling confrontation and criticism. Because they have a feeling for both their strengths and their weaknesses in interpersonal living, they usually aren't caught off guard when they are confronted. Let's consider an example.

> *Rob:* "I don't think that I'm getting anything out of these sessions. I'm just the same as I was the day this group started. I don't see any big differences in my relationships with people outside the group. I doubt whether any of you do. I should probably leave and not waste my time any more."

> *Sally:* "I think you *are* wasting time. You really don't put yourself into the work here, so what can you expect outside—a miracle?"

> *Ray:* "There's been no payoff for you—just dreary work and no results. I'm not sure whether you want to find out why there's been so little payoff."

Here Sally is defensive (responding by an attack), while Ray tries to understand Rob, making his response tentative but clear and direct. A genuine person is at home with himself or herself, though not in a smug way, and as a result can respond to criticism with understanding and a willingness to explore negative feelings. A genuine person can say to himself or herself "Am I doing something that's contributing to this person's negative experience?"

185

4. *Consistency among thoughts, feelings, and behaviors.* A genuine person avoids discrepancies between what he or she thinks or feels and what he or she actually does or says. Often a lack of genuineness results in inconsistent behaviors. Examples:

- Tom feels very angry with the other members of the group but smiles a great deal and says nothing about his anger.
- Trudy thinks that Jerry is being unfair in his criticism of Mike but remains silent.
- George says that he wants to reveal more about himself in the group but never gets around to it.
- Yolanda is very enthusiastic about the group one week, but she comes late and sits around in a distracted way the next.
- Ted says that caring is one of his most important values, but he seldom responds to anyone in the group with accurate understanding.

Good communicators avoid these kinds of discrepancies, for they tend to make interactions phony.

5. *Openness.* A genuine person is open and is capable of deep self-disclosure. At the same time, such a person isn't wildly open, since, as we've seen, self-disclosure isn't an end in itself. But he or she does feel free to reveal himself or herself intimately when intimacy is called for. Because you know where a genuine person stands, you can trust him or her.

It's impossible to practice genuineness by itself, since genuineness is a set of attitudes and behaviors that influence the entire communication process. However, it's important that you receive feedback from your fellow group members on the genuineness of your interactions with them. It will help to make feedback concrete if group members keep clearly in mind the *behaviors* that indicate genuineness. Here is a checklist of these behaviors and of some common nongenuine behaviors.

A Checklist on the Communication of Genuineness

- Is the communicator natural, or does he or she have a "group self" and a "real self"?
- Does the communicator get stuck in certain roles, such as

parent, helper, client, question asker, sympathizer, rescuer, organizer, or others?

- Is the communicator becoming more at home and relaxed with the skills being learned, or does the person still sound as if he or she were practicing?
- Is the communicator spontaneous, or are there ways in which he or she is rigid (always acting the same way in every group session, saying things that sound planned, always waiting for the "right" moment to say something, and so on)?
- Is the communicator defensive, or is he or she open to dealing with negative criticism and even attack?
- Does the communicator avoid attacking others?
- Is the communicator open, sharing himself or herself willingly when it's appropriate? Does he or she always have to be asked to share?
- Is the communicator tactful and caring but still direct in his or her communication? Do you know where he or she stands with you?
- Is the communicator one person at one time and another later on, or is he or she generally consistent? Can you count on this person?
- Does the communicator play games? If so, what games? What payoff does he or she get from playing games?
- Does the communicator try to manipulate or control others? If so, how does he or she try to do it?

If you tell a person that he or she either is or is not genuine, you should be able to back up your statement with concrete behavioral examples.

Respect: A Behavioral Approach

Respect is such a fundamental notion that you might have some difficulty defining it. Respect is a way of looking at people (it even comes from a Latin word meaning "to look at"). It means appreciating people simply because they're human beings. It implies that being human is an important value in itself. In the case of respect, actions literally speak louder than words. You'll probably never say to one of your fellow group members "I respect you because you are a human being." Respect is communicated to others by the ways in which you act toward them. What are some of the attitudes and behaviors that indicate respect?

1. *The attitude of being "for" others.* Sometimes, when I meet people, I feel accepted almost immediately. I don't have to prove myself in some way before being accepted. This I take as respect. At other times, however, the opposite happens. I feel that people are somewhat suspicious of me, that I have to prove myself. Just being a human being isn't enough; I have to be strong or witty or smart or conservative or liberal or something else in order to gain acceptance. I feel a basic lack of respect.

Being basically "for" others doesn't mean not caring about what they do or whether they grow. Parents can be "for" their children and still place demands on them. In fact, respect in its highest form *is* demanding: it means that you expect another to be as fully human as possible. This highest level of respect, however, belongs in the closest relationships. We'll consider this kind of respect when we discuss the skills of challenging.

2. *Attending.* Attending is itself an excellent way of showing respect to others. Attending behavior says, in effect, "I'm with you. I want to be with you. I'm committed to working with you. It's worth my time and effort to be with you." It's easy to tell the difference between a person who is "there" psychologically and a person who isn't.

3. *Cooperation.* Cooperating with others is a sign of respect. Your training group is basically a learning community, and respect means a willingness on your part to give yourself actively to the work of the group. Learning communication skills, reviewing your interpersonal style, establishing relationships with your fellow group members, and making attempts to change some of your interpersonal behaviors all amount to a lot of work and take quite a bit of self-investment. Your work is an indication of your respect both for yourself and for the others.

4. *Regard for the uniqueness of each person.* The fact that all of the people in the group have common goals doesn't mean that everyone has to do everything in the same way. Each person has different strengths and different resources and uses interpersonal skills in different combinations. The most obvious lack of respect would be wanting everyone to act just like you. You might not like certain interpersonal styles—for instance, you might not like it when someone spends a great deal of time understanding others and very

188

little time challenging them—but you can still understand and respect different styles.

5. *Seeing others as capable of taking care of themselves.* Respect also means respecting the resources that others have. Each of the people in the group has an obligation to develop his or her own resources. It's respectful to encourage others to use their resources; it's disrespectful to try to take over their lives or to become a helper instead of a helpful equal. Many of your resources and those of your fellow group members may be blocked in various ways (for instance, you might have learned from childhood to fear intimacy with others) or may be underdeveloped (perhaps you've never been challenged to use your interpersonal resources). To help one another free up and cultivate resources isn't the same as taking over one another's lives. Ultimately, if a participant chooses to live a less effective interpersonal life than he or she is capable of living, even this choice is to be respected.

6. *Assuming the other person's good will.* Respect calls for acting on the assumption that the members of the group are serious about bettering their interpersonal lives. This should be the assumption made of everyone in the group until it proves false. Even when a group member acts in a negative way, a good first impulse is to try to understand what's happening rather than to judge that he or she has ill will. You might also ask yourself whether you are contributing to the other person's negative attitude or behavior. Especially in the early stages of the group, respect takes the form of suspending critical judgment of others.

On the other hand, having respect doesn't mean that you simply don't care whether another person grows or not. Since all the participants are in the group to grow, respect can be both understanding *and* appropriately demanding, even from the beginning of the group. Showing respect includes making the assumption that the members of the group are there freely and that they are willing, however gradually, to pay the price necessary for interpersonal growth. One way of looking at the "demand" dimension of respect is this. The other person's being in the group can be taken as an indication that he or she wants to place some demands on himself or herself. If the group members help one another meet each person's own demands on himself or herself, then they can be demanding at the same time as they respect each person as self-determining.

7. *Understanding.* One of the best ways of showing respect is by working to understand the other members of the group. I know that a person respects me if he or she spends time and energy trying

to understand me. All of the behaviors associated with the communication of understanding, then, are behaviors that indicate respect.

8. *Warmth.* Warmth is another interpersonal characteristic that is difficult to define. In one sense, warmth is the physical expression of understanding and caring. Ordinarily warmth is expressed nonverbally through gestures, posture (closeness, touching), tone of voice, and facial expression. *Appropriate* warmth is one way of showing respect, but it's not the most important way. Being appropriately warm in the beginning of a relationship helps to let the other person know that you are basically *for* him or her, but some people are "warm" all the time. They never want to do anything that might disturb another person. Warmth, in this case, becomes a form of *disrespect.* Let's consider an example of inappropriate warmth.

> *Curtis:* "I'm too easy with myself. As you can see, I've let myself go physically. And I don't read much any more, which means that my conversations can be pretty dull. I take my dissatisfaction with myself out on other people—I think that's one of the reasons I fight so much with people here."

> *Lee* (in a very warm tone of voice): "First of all, Curtis, men your age tend to get a little paunchy. You don't look so bad to me. And I find you an interesting person. I think you might be too hard on yourself."

> *Donna:* "Curtis, I hadn't realized that your manner of reacting here was associated with your physical picture of yourself. That helps me understand you better. I need to talk this out further, too."

Here Lee's warmth is misplaced. He goes so far as to suggest that Curtis lower his standards. On the other hand, Donna's response demonstrates that she respects Curtis for his self-disclosure and appreciates the chance to face the issue he brings up. Her response also takes some of the pressure off Curtis by showing that it will help her to discuss the matter further. It rewards Curtis for his self-disclosure, but it doesn't downplay what he has said.

9. *Support and encouragement.* A climate of support arises partly from the encouragement that group members show one another. When a person succeeds in placing demands on himself or herself in order to grow, others' recognition of his or her success is encouraging and rewarding.

190

Audrey: "I find myself taking more time to listen to others and to respond with understanding—not only here but outside the group as well. This has made me much less tense with people. I didn't realize that such a simple thing could improve my relationships so much."

Larry: "If you learn how to confront others more reasonably now, I'll bet you'll find your relationships even more rewarding."

Sarah: "I do see you much improved in your skills and certainly much more relaxed. It's a lot easier for me to be with you now. And it sounds like all of this is pretty rewarding for you."

Larry ignores what Audrey has already accomplished, whereas Sarah's recognition of Audrey's success is encouraging.

10. *Genuineness as respect.* Being genuine in your relationships with others is, of course, a way of showing respect. Therefore, the behaviors listed in the checklist on genuineness are also ways of showing respect for others.

A Checklist on the Communication of Respect

Respect, like genuineness, is impossible to practice by itself. However, you can get feedback from others on whether your behaviors and attitudes communicate respect. This checklist should help you and the other group members to give one another that feedback.

- Does the communicator seem to be "for" others in a caring but unsentimental way?
- Is he or she *working* at communicating with the other members of the group?
- Does the communicator avoid being judgmental?
- Does the communicator respond with accurate understanding frequently and effectively?

191

- Is the communicator appropriately warm? Does he or she avoid coldness, aloofness, "canned" warmth, inappropriately intimate displays of warmth?
- Does the communicator recognize others' resources, strengths, and successes?
- Does the communicator give his or her full attention to the person speaking? Does the communicator's nonverbal language tell the speaker "I'm with you"?
- Does the communicator avoid behavior that would indicate an attempt to control or manipulate others?
- Does the communicator treat different people differently (that is, does he or she respect the uniqueness of each individual)?
- Does the communicator give others a chance to live up to the contract, to learn the skills, to change?

If you tell a person that he or she is or is not showing respect, you should be able to back up your statement by pointing out concrete examples of behaviors.

The Skills of Challenging

You've so far been introduced to two major sets of skills—the skills of letting yourself be known (self-disclosure, the expression of emotion, and concreteness) and the skills of responding to others (accurate understanding, genuineness, and respect). These two sets of skills provide the basis for a climate of support in the group and in human relationships generally. Through self-disclosure, you risk letting yourself be known. And, by responding humanly to those who disclose themselves to you, you justify the risk they take in disclosing themselves.

Since I've used the word *support* fairly often in these pages, it might be useful to get a clear picture of what I mean by it. There is a climate of support in a group if:

- the group members trust one another and feel that they are trusted.

- the group members take the time to communicate understanding to one another.
- the group members respect one another and feel that they are respected.
- the group members can make mistakes, and the others help them learn from their mistakes.
- the group members feel at home with one another.
- the group members take the work of the group seriously.
- the group members cooperate with one another.
- the group members appreciate one another's strengths.
- the group members feel that they can take reasonable risks, and they see others taking reasonable risks.
- the group members have clear, reasonable goals and are working to achieve these goals.
- the group members not only understand one another but are willing to challenge one another.

As we begin to consider the skills of challenging, it's important to remember that challenging is most effective in a group in which there is a great deal of support. This third major set of skills can be strong medicine in human relationships. Strong medicine can heal if it's used properly, or it can have destructive effects if it's used carelessly. In the same way, the skills of challenging can deepen, enrich, and intensify human relationships if they are used properly. If misused, however, these skills can hurt others and destroy relationships. In the deepest and richest human relationships, people both support and challenge one another. They challenge one another to pursue goals and values that each has freely chosen. Deep support plus caring challenge is a rich combination in human relationships.

If both parties in an intimate relationship have to learn how to use these skills well, then one will not end up as a helper and the other as a client. As I've said before, friends or equals help one another without getting locked into the roles of helper and client. Therefore, this group experience will stress *mutuality* in the use of challenging skills. The skills will be based on your ability to disclose yourself and especially on your ability to respond to others with understanding. If you haven't learned the skills of responding well, you're not yet ready to learn the skills of challenging. *A person who responds poorly to others will also challenge others irresponsibly or ineffectively. A person who doesn't first understand others doesn't have the right to challenge them.*

193

A Pre-Challenging Skill: Identifying Other People's Strong Points

Your understanding of others is complete if it also includes an understanding and appreciation of their good points, resources, abilities, and strengths. In addition, it's most helpful to get feedback yourself on your good points. Feedback from others can help you appreciate your own resources more and motivate you to further develop your abilities. Many people fail to appreciate themselves because they keep looking at what they do poorly instead of at what they do well. They keep looking at what they can't do instead of learning to appreciate the strengths and abilities they're not using fully.

The ability to identify strengths in others is extremely important if you're going to challenge them. A person will be much more likely to accept feedback from you telling him or her what he or she fails to do if you first show this person some appreciation of his or her strengths. Moreover, the use of challenging skills assumes a good relationship between you and the person you're challenging. Before challenging someone, you should ask yourself whether your relationship with that person is a good one. People react negatively to challenges from those who don't know them, who dislike them, who don't understand them, or who are indifferent to them. I will repeat this caution later on. However, at this point you might review the quality of the relationships you've established with your fellow group members. From whom are you willing to receive some form of challenge? Whom do you think you have earned the right to challenge?

In the next three chapters I will consider three skills of challenging: deeper understanding, confrontation, and you-me talk (immediacy). As you will see, these skills are interrelated. Much of the challenging that goes on in everyday life is done poorly. These chapters provide a blueprint on how to challenge responsibly, caringly, and effectively.

Exercise 34
Identifying Strengths in Others

The purpose of this exercise is to make you more aware of your own interpersonal strengths or resources and to help you give others feedback on their interpersonal strengths and resources. A person who sees his or her own strengths may also see that he or she has the resources to move forward when challenged.

Directions

a. Identify three interpersonal strengths or good points that you think you have.
b. Do the same for each of your fellow group members. Write them down briefly.
c. In identifying strengths, be as concrete as possible. Use examples.
d. Your instructor will tell you how to share what you've written with your fellow group members.

Example

"I see Raul as interpersonally strong. By that I mean that he's very active in the group. He initiates conversations with others. He reveals himself deeply. He has talked about his fears about his sexuality. And yet he doesn't seem to get very anxious when he does these things. All of this strikes me as strength. This strength of his has two effects on me. It makes me like being in this group with Raul. He adds some excitement to the group. It also makes me a little nervous, because what he does challenges me. His risking of himself invites me to risk myself."

How does the feedback you receive from others agree with what you've written about yourself? Do others see you (with respect to your strong points) as you see yourself? What strengths do others see in you that you don't see in yourself?

Receiving feedback on your strengths is, in a way, challenging in itself. It gives you a positive picture of yourself, but it also encourages you to keep living up to this picture.

Deeper Understanding

9

Others often see us and understand our behavior in ways that we don't see or understand ourselves. Sharing this kind of information with one another can enrich and deepen our relationships. This chapter describes this kind of deeper understanding and suggests ways of sharing it with one another.

The Difference between Basic Understanding and Deeper Understanding

One kind of challenging is letting other people know something you see about them that they themselves don't see, at least not clearly. For instance, it might happen that someone says to you "Sometimes I wonder whether you really like yourself—you're always giving yourself the dirty end of the stick," and a light goes on inside you. Your friend has put into words something that you've been feeling or half suspecting. You might say to yourself "Come to think of it, there *are* times when I don't seem to care very much about myself. Sometimes I think that I'm mean to people just so they'll be mean to me back." In this case, what your friend has done is communicated to you what I'll call *deeper understanding.* What I mean by *deeper* understanding will become clearer if it is distinguished from the kind of *basic* understanding discussed in Chapter 7.

As we've seen, responding with accurate understanding means letting another person know that, to some degree, you see his or her world in the same way that he or she sees it. You have a *feeling for* the other person's world, and you communicate this "feeling for" to the other person.

Ivan: "Last night I found out that we're going out on strike tomorrow. My wife and I had a fight this morning. And then I got a call that my son broke his leg playing baseball and was in the hospital."

Louise: "You must be feeling awful. Your world is collapsing on you."

Ivan: "You're damn right it is!"

In this exchange, Louise isn't telling Ivan anything he doesn't already know. Rather, she's letting him know that she *understands* what he's going through. When he says "You're damn right it is," he lets her know that she has hit the mark.

Deeper understanding is different from this kind of basic understanding. Through deeper understanding you communicate to another person what he or she does *not* see or experience very clearly. You might immediately ask "But how can *I* see what the other person doesn't see?" Suppose that you and I are in a training group together. If you work at understanding and "being with" me, you soon begin to know me in deeper ways. In fact, you can come to know me in ways in which I can't know myself directly. After all, even though I'm living with myself constantly, I keep seeing myself *from the inside.* You see me *as others see me.* Therefore, you're a potential source of useful information about me. You can see things about me that I'm not looking at or that I'm blind to. If you care about me and pay a great deal of attention to me, and if, in addition, you've spent time communicating accurate understanding to me, then you're in an excellent position to cue me in on some of my blind spots. When you do this, you communicate *deeper understanding* to me. Consider the following example.

Jim: "I don't know what's happening to me in this training group. I think I try just as hard as everyone else, but I still don't feel like part of the group. I don't seem to fit at all. I don't form relationships as easily as the rest of you. It's probably my own fault, but all my work seems to go down the drain. But I don't know what else I can do."

One possibility is to respond to Jim with basic, accurate understanding.

> *Frances:* "It's frustrating and depressing. You put in as much effort as everyone else, but it doesn't seem to pay off for you."

Frances communicates to Jim that she has a basic understanding of Jim's world. However, let's assume that this isn't the first time that Jim has compared himself unfavorably to the other members of the group. By this time, Frances has a hunch that something else is happening to Jim that Jim himself doesn't see or at least doesn't see clearly. So Frances *plays a hunch,* based on her attending to and responding to Jim up to this point in the group.

> *Frances:* "It's depressing to put in so much effort and still feel that you're not getting anywhere. *It almost sounds to me as if you're beginning to feel a little bit sorry for yourself and that that might be making things seem even more impossible.*"

The italicized words constitute deeper understanding. This understanding goes beyond Jim's world as Jim sees it. It gets at something that Jim doesn't see or that he isn't yet fully aware of. Notice that deeper understanding isn't based on any single statement that Jim makes about himself. Rather, it's based on a *pattern* that Jim is beginning to show. Notice also that Frances has initiated her hunch with the qualification "It almost sounds to me," as if to check the accuracy of her perception with Jim. This qualification gives Jim the chance to say "No, that's not it; rather it's . . . ."

Frances' deeper understanding of Jim is possible only because she has been attending carefully to Jim, all along communicating basic understanding to him. As is true of basic understanding, however, deeper understanding must be *accurate* to be effective. It might take Jim a while to see that he has begun to trip over his own self-pity, because he hasn't had a clear picture of himself in this respect. But, if Frances is on the mark and communicates her deeper understanding carefully, so that it doesn't sound like an accusation, Jim might well respond as follows:

> *Jim* (after pausing for a while): "You know, I think you might be right. It's not so much the work here. I'm beginning to stumble over myself. I've had these self-defeating attitudes before."

Jim begins to see the light. He doesn't feel judged. Instead he gets in touch with himself in a new way, and his new understanding will, hopefully, help him pull out of his self-pity.

Deeper understanding as hunch or intuition. No doubt you sometimes have *hunches* about what's going on inside people you know. Your hunches are probably based on a great deal of experience with these people. They aren't psychological magic; rather, they're a part of being socially intelligent. If you communicate these hunches caringly and skillfully, often you can help others to get more in touch with feelings or issues that aren't very clear to them.

> *Clifford:* "I'm really disappointed that I didn't get that job. I had really counted on it. I could be saving money, and that would make it a lot easier for Nora and me to get married next year."

> *Elaine:* "Clifford, I can see the disappointment written on your face, but if I were you I'd be angry, too. From what you've told me before, you got a raw deal."

There can be a huge difference between what people actually *do* feel and what you think they *should* feel. Evidently Elaine has seen signs of suppressed anger on Clifford's part. Elaine also assumes that it might be good if he could just admit his anger and get it out in the open, instead of letting it eat away at him. Notice, however, that Elaine doesn't tell Clifford that he *should* be angry and that he *has to* admit it. She approaches the whole issue of his anger carefully, *tentatively.* The words "but if I were you" make her statement tentative. These words keep what she says in the area of hunch or intuition instead of absolute statement or accusation. I'll have more to say later on about being tentative in your use of challenging skills.

Deeper understanding is useful only to the degree that it's accurate; otherwise it *does* sound like an accusation. Furthermore, if you either use too much deeper understanding or use it poorly, you'll begin to play psychologist and run the risk of destroying mutuality between yourself and others. There are a number of ways in which you can develop intelligent hunches that can be communicated to others as deeper understanding. Let's review some of these.

1. *Expressing what the other person only implies.* Putting into words something the other person only implies is the most basic form of deeper understanding. Very often there is both a *surface meaning*

200

and a *deeper meaning* to what people say. Deeper understanding gets below the surface to what is usually not stated clearly, directly, and openly. The deeper meaning is usually hidden, though not necessarily on purpose, in signs or cues, either verbal or nonverbal. Let's consider an example.

> *Howard:* "I've gotten in touch with an ability to relate to others that I didn't even know I had. I see that I'm caring, that I can talk concretely, that I'm not afraid to reveal myself to others who give me half a chance. I'm not trying to blow my own horn. I'm just saying that these discoveries are important to me."

It's possible to respond to Howard with basic understanding.

> *Rachel:* "These interpersonal abilities are very real, and their discovery has been, well, exciting for you."

But let's suppose that Howard has been a person who started out slowly in the group and who is now working very hard. The amount of effort that he puts into the group is one of the cues that can be used to interpret his statement. Thus Carmen, who has come to know Howard quite well, responds:

> "I hear a note of real determination in your voice. Now that you've gotten in touch with these abilities, you're going to make them a part of your life. And that's probably even more exciting than just doing well in the group."

Carmen goes deeper than Rachel and reads—hopefully, in an accurate way—what Howard hasn't said directly but has *implied* by his actions. In a sense, then, deeper understanding is a kind of interpretation—not in the sense of psyching out the other person but in the sense of seeing his or her statements *as part of a bigger picture.* In this case Carmen sees Howard's remark in the wider context of his behavior in the group. If Carmen is accurate, and if Howard is ready to receive this kind of deeper understanding, then he will respond positively:

> "Yeah, I'm not implying that I'm a totally new person or anything like that. But what I'm learning *is* making a difference in my life. I think you're right. I want to get everything I can out of this group."

201

If instead Howard became confused or denied what Carmen had said, this would be a sign that she hadn't been accurate or that Howard wasn't yet ready for that particular bit of deeper understanding.

The readiness of a person to receive and listen to deeper understanding is very important. Communicating deeper understanding is a skill of challenging because it asks a person to take a deeper look at himself or herself. Sometimes taking a deeper look at oneself can be an unpleasant task. Through the communication of deeper understanding, you and your fellow group members place demands on one another to take a deeper look at your interpersonal styles. As I've already implied, the best way to help one another become ready to take this deeper look is to communicate genuineness, respect, and basic understanding to one another from the very beginning of the group. If you and the others in your group have been responding to one another in the ways described in Chapters 6, 7, and 8, then trust has most likely developed among you, and you're beginning to be ready for the stronger medicine of deeper understanding.

2. *Identifying themes.* As you and your fellow group members go about the business of establishing and developing relationships with one another, certain themes in your interpersonal styles probably will come to light. For instance, as you explore your interpersonal behavior outside the group and get feedback from others inside the group, it might become clear that you have a tendency to control others. Perhaps, if you don't like what's going on in the group, you say something like "Hey, I don't think we're getting anywhere," when what you really mean is that you're not getting what you want in the group. This need to control others in order to get what you want emerges as a *theme*, a pattern, in your interpersonal style.

If you begin to see a theme emerge in someone else, a theme that he or she doesn't yet see clearly, then it is a communication of deeper understanding to give this person feedback on what you see. For instance, you might see a theme of dependence emerge as one of your fellow participants interacts with others in the group.

> *You:* "Sean, you and I get along quite well here. At least that's my evaluation of how we stand with each other. But I'd like to describe a few things that I see about you and get your reaction. One is that you're very hesitant to challenge anyone here, perhaps especially the people you like the most. Another is that you listen very carefully when Fred [the instructor] is talking. It seems that you almost always agree with him. It's almost as if you're *too* respectful. You've asked me several times outside the group to give you feedback on how

you're doing. The message I'm beginning to get from all this is: 'I'm not entirely my own person. I depend on others quite a bit. I don't value myself first; I wait until I see how others value me.' Perhaps all of this is too strong. These are really impressions I'd like to check out with you."

Sean: "What you're saying is very painful for me to hear. It's even more painful because it's right."

The themes that begin to appear in what a person has to say about his or her interpersonal style may refer to emotions, experiences, or behaviors.

• *Emotions*. Some themes that concern emotion are: having a tendency to give in to self-pity; being an enthusiastic, optimistic person; being a fearful person in interpersonal situations; being an affectionate, warm person; having a tendency to get depressed easily; being very self-confident. It's useful to get feedback from others with respect to your emotional tone and style. You may not realize that you might come across as a moody or suspicious person, and yet how you come across is affecting how people react to you all the time.

• *Experiences*. Some themes that refer to interpersonal experiences are: being a victim, being loved, being seduced (sexually or otherwise), being feared, being the object of rescuing attempts, being looked to for leadership, and others. Remember that experiences refer to what happens to you rather than to what you do. However, if people, using deeper understanding, let you know that others are always trying to take care of you, you might begin to ask yourself "What is it in me, what do I do, that makes other people want to take care of me? What signals of helplessness am I sending out?" Or if others look to you for leadership, you can begin to ask yourself "What resources do I have—resources that I don't fully appreciate—that make others look to me for leadership?"

• *Behaviors*. Interpersonal-behavior themes might include: caring for others, controlling others, being aggressive, being assertive, avoiding intimacy, being cooperative, blaming others for what goes wrong with you, and others.

You can't see yourself as others see you unless you get straightforward feedback from others. Getting feedback on your emotional, experiential, and behavioral themes is one way of getting to see yourself as others see you.

3. *Pointing out logical conclusions.* Sometimes we don't draw the logical conclusions of what we're saying. So another way of getting at deeper understanding and communicating it to another per-

son is to point out the logical conclusions of what he or she says. Let's consider an example.

> *Rudy:* "I know I'm quiet here, or at least relatively quiet. But I don't think that's any reason for people to pick on me or to make me feel like a second-class citizen. I've got a lot going for me. I could say a lot of the things that other people say here, but they move in first."

> *Nate:* "The logic of what you're saying seems to be 'One, I'm not a second-class citizen, because I've got interpersonal strengths just like the others here; two, I'm going to begin to take my rightful place in this group. I'm not going to let others squeeze me out.' That's what I think you're saying, but I'd like to check it out with you."

Nate points out a logical conclusion to what Rudy has said. Rudy didn't directly *say* that he's going to act in a way that will show he isn't a second-class citizen, but that seems to be what he means. Nate's hunch or guess, based on what Rudy has said, is that Rudy wants to be more assertive; so Nate communicates that hunch to Rudy as deeper understanding. But remember that a guess or a hunch is useful only to the degree that it's accurate.

The Importance of Mutuality in Communicating Deeper Understanding

The communication of deeper understanding is a skill of challenging because it makes us think more deeply about ourselves. It makes us examine what we really mean by what we're saying, what we're doing, and what we're not doing. However, if you begin to specialize in the communication of deeper understanding, always telling others what your guesses and hunches are about their behavior and their personalities, then you begin to play psychologist. Since challenging skills are "stronger medicine," specializing in challenging skills is more likely to turn you into an amateur psychologist than is specializing in the communication of basic understanding. One way to avoid playing psychologist in the group is to share with others some of the hunches you have about yourself and your own behaviors. For instance:

204

"I've begun to notice that people here tend to say 'Thank you' to me after I've talked to them. I'm beginning to suspect that I'm coming across like a parent—a nice parent, but still a parent. Could you give me some feedback on that? I don't want to be parental or a helper."

Someone might then point out to you that you respond very well to others, usually with accurate understanding, but that you don't reveal very much about yourself. That pattern does make you seem like a nice parent or a warm helper. Then you can begin to do something about changing this pattern. As the group moves forward, you and your fellow group members will begin to use more and more of your challenging skills, but it's important that you do so mutually. Good groups are characterized by *mutual* support and *mutual* challenge.

Being tentative in the wording of challenges. Very often it's not an easy thing to be challenged. The usual tendency of a person who is challenged is to react as if he or she had been attacked—to become defensive or to attack back. You can help reduce this tendency by the way in which you word your challenges. If there's something of a guess or a hunch involved in many of the challenges we give one another, then a challenge need not be presented as an absolute statement that can't be wrong. Challenges aren't ways of saying "I'm right, and you're wrong"; nor are they ways of controlling or manipulating one another. Challenges are ways of "being with" one another more intimately, more intensely. In the examples of deeper understanding used in this book, there are certain qualifications that make them *tentative*. For instance:

> *Carlos:* "You've already said that you know you're quiet, Rudy, more quiet than you want to be in here. *If I'm not mistaken,* there is *at times* something more there than just being quiet. I see you as moody *at times.* It's partly that you're quiet, but it's also that you give some nonverbal signs that you're experiencing some negative emotions. *Does what I'm saying click with you at all?*"

The italicized words qualify the communication of Carlos' deeper understanding of Rudy. They make it *tentative*. They keep it from sounding as if Carlos were dumping on Rudy, attacking or accusing him.

Some people see tentativeness as a lack of genuineness. They say that you're sugarcoating your feedback instead of giving it to them straight. But there's a big difference, I believe, between making a response tentative and sugarcoating it. Sugarcoating is a lack of respect for the other person. It says that he or she is weak and can't hear things directly. Sugarcoating says that you have to take care of the other person. On the other hand, being tentative means that you don't see yourself as faultless or infallible. It also means that you realize that it isn't easy for others to hear challenges. You can be careful of another person without "taking care of" him or her. One good way of being tentative is to do some perception checking with the person you're challenging. In the last example, Carlos ends his response with a perception-checking question. In the examples of challenging that follow in the next two chapters, notice the ways in which tentativeness is expressed, and discuss with others whether this comes across as sugarcoating or as genuine tentativeness.

Exercise 35
Deeper Understanding:
An Exercise in Self-Exploration

One way of making sure that you use challenging skills carefully is to use them on yourself first. The purpose of this exercise is to help you think more deeply about some aspects of your own interpersonal style. In a sense, it's an exercise by which you communicate deeper understanding *to yourself*. The exercise asks you to look at yourself at two different levels—one corresponding to basic understanding, the other corresponding to deeper understanding.

Directions

a. Choose an area of your interpersonal style—a pattern, a relationship, your feelings about yourself as a person who relates to others—that you'd like to examine more deeply. Choose an area that you would be willing to share with the other members of your group.
b. Briefly describe the area on which you're going to work.
c. As in the examples that follow, first write a statement that reflects *basic understanding* of the area you've chosen; then write a statement that probes deeper—one that reflects a *deeper understanding* of the chosen area.

Example 1

- *Area to be explored:* "I'm concerned about the way I am 'with' people in social or friendship situations."
- *Basic, accurate understanding:* "I enjoy being with people. I meet people easily, and I'm generally well liked. I make others feel at home. I appear to be quite outgoing—I'm humorous without being a 'funny man.' I try to understand other people's points of view, and I show interest in what others are doing. I'm careful with others; I don't dump my negative feelings on people, and I'm slow to confront them."
- *Deeper understanding:* "When I'm with others, even though I appear to be outgoing, I'm not 'all there.' I don't share my deeper feelings. There's something almost superficial (maybe that's too strong a word) about my 'being with' others. I hold back. In my deepest moments, I'm alone with myself, and that's the way I like it. People like me, but they don't get very close to me, even though sometimes they think they do. I haven't learned to share my deeper self with anyone, and I'm not sure whether I want to. I may even be afraid to."

Example 2

- *Area to be explored:* "I've begun to wonder just how deeply I'm committed to my human-relations-training group."
- *Basic, accurate understanding:* "I'm an active group member. I enjoy the work of establishing relationships with others. I think that others see me as open. I spend a fair amount of time communicating basic understanding to others. I never get challenged on the quality of my participation. I'm technically good at most of the skills we've learned, including skills of challenging. Most people in the group have told me that they think I'd make a fine trainer."
- *Deeper understanding:* "On deeper reflection, there are ways in which I'm dissatisfied with my participation in the group. First of all, I think that in many small ways I control what happens in the group. I'm 'nice,' and so others don't challenge me much. I share myself, but not too deeply. In many ways I see others saying 'Let's not make this group too intense,' and I don't say anything. It may be that I'm an excellent member in a group that's not very intense. Some-

207

times I feel phony, but I get rid of those feelings as quickly as I can. All of this makes me wonder how intensely I want to live my interpersonal life."

Now write out basic-understanding and deeper-understanding statements about four interpersonal and/or group issues that are important to you. Be as concrete as you can in both the basic-understanding and the deeper-understanding statements. Share your statements about yourself in whatever way is determined by your instructor. Get feedback from your fellow group members. Does your deeper understanding of yourself surprise them in any way? In what ways?

Exercise 36
Communicating Deeper Understanding to Others

Before trying to communicate deeper understanding in your group or in your relationships outside the group, you can, through this exercise, prepare yourself for using this skill.

Directions

These directions are the same as those in the preceding exercise, except that your attention is now directed toward your fellow group members instead of toward yourself.

a. Consider one of your fellow group members. Choose some aspect of this person's interpersonal style that you'd like to consider, both in terms of basic understanding and in terms of deeper understanding. Remember that you're trying to communicate understanding; you're not psyching the other person out.

b. Briefly describe the area or issue you want to consider.

c. As in the following example, first write a statement that reflects *basic understanding* of the area you've chosen. Write the statement as if you were speaking directly to the person. Then write a second statement that reflects a *deeper understanding* of the person. Again, write the statement as if you were speaking directly to the person.

d. Repeat the exercise for each person in your group.

Example

- *Area to be explored:* Ingrid is honest in admitting her inter-personal defects, but honesty doesn't seem to be enough for her.
- *Basic, accurate understanding:* "Ingrid, you share your weaknesses and your fears with the other members of the group quite freely. You even wonder whether it might be useful to see somebody for counseling. Your honesty here seems to put you at ease; it gives you a kind of freedom in the group that you appreciate. You get a great deal of re-spect for being honest, and this seems to buoy you up. Despite your disclosure of weaknesses, you're accepted here, and I believe that you feel this acceptance deeply."
- *Deeper understanding:* "Ingrid, when I look at you more closely, I see you fidgeting when people let you know that they accept you. I think that you feel that it's great to be accepted but that you also feel you want more than that here. It's almost as if you're saying to yourself 'Honesty isn't enough.' I think that you want to begin to change some of your behavior. For instance, you want to be more assertive. And yet no one is challenging you to become what you want to be. My bet is that at times you feel disappointed that few, if any, demands are placed on you. I think that you may be afraid that you'll leave this group much the same as when you entered it."

In writing out your statements of basic understanding and of deeper understanding, try to be as concrete as possible. Also, try not to *accuse* your fellow group members. Be appropri-ately tentative in how you word your deeper understanding. Finally, share your deeper understanding with one another in whatever way is determined by your instructor or trainer.

Confrontation

This chapter explains what is meant by "confrontation" and suggests how you can both confront others in a positive, caring, and responsible way and receive confrontation in ways that can help you grow interpersonally.

Confrontation: What It Is

Confrontation is a word that strikes fear in many people, because to them it connotes the same thing as *attack*. If you've been the victim of irresponsible and poorly delivered confrontation, you might be somewhat fearful yourself. Some people say that we should get rid of interpersonal confrontation altogether, but, for better or for worse, it seems to be a permanent part of human living. Yet confrontation can be a very growthful kind of communication. It can actually lead to greater intimacy and fuller interpersonal living—*if* it is done well. There isn't a great deal of evidence that many people do it well.

Confrontation as invitation. As I am using the word, *confrontation* means anything that invites a person to examine his or her interpersonal style—emotions, experiences, and behaviors—and its consequences (for instance, how it affects others) more carefully. The communication of deeper understanding is confrontational in this sense, because it digs deeper than does basic understanding and asks the other person to look at himself or herself more carefully. Through confrontation we challenge the discrepancies, the distortions, the games, the unwillingness to understand, the unproductive behaviors that plague our interpersonal lives. Two psychologists (Berenson and Mitchell, 1974) have described five different

211

kinds of confrontation. Let's examine these types of confrontation as a way of discussing *what* to confront in others.

1. *Information confrontation.* Sometimes we act in unproductive ways with one another because we lack some kind of information. If you give one of your fellow group members information that he or she either doesn't seem to have or doesn't seem to grasp fully—information that the person needs to involve himself or herself more effectively with others—then you're engaging in an information confrontation. Let's take an example.

> *Helen:* "Jackie, there's one part of the group contract that I'm not sure you're aware of. I'd like to check it out with you. You talk a lot about your interpersonal life outside the group. For instance, you've talked a great deal about how dependent you've been on your parents. You've talked about how that affects you with your boy friend. We deal with these things as problems 'out there,' but neither you nor the rest of us do very much to relate what you say to your relationships here in the group. And it turns into a kind of counseling session. The here and now goes out the window. Does this make any sense to you?"

> *Jackie:* "Now that you bring it up, I do remember that part of the contract, and I guess I have let it slide. Yes, I do end up in a counseling session, with most of you as my counselors. I'd like to talk about how I can change that."

In this confrontation, Jackie is given information that has slipped her mind. She recognizes the truth of what Helen is saying and calls on the resources of the group to help her change her manner of participation. If members of the group consistently violate some provision of the contract, this kind of information confrontation is useful to get at what is underlying the participants' ignoring of the contract.

2. *Experiential confrontation.* As you involve yourself more deeply with the other members of the group, you'll notice that at times you *experience* another member *differently* from the way in which he or she experiences himself or herself. There's some *discrepancy* between the way you see that member and the way that member sees himself or herself. Let's take a look at an example.

> *Max:* "Ginny, a number of times you've said here that you're unattractive. And yet I know that you get asked out a lot by

men. And I see people here in the group saying 'I like you' in various ways. I'm having a hard time putting this together with your calling yourself unattractive."

Ginny: "What you're saying helps me clarify what I mean. First of all, I'm certainly no raving beauty. When others say that they find me attractive, I think they're talking about my personality. I do care about people. But there are times—and I feel ashamed even saying this—when I wish I were more physically attractive. And, more important, many times I *feel* unattractive. And sometimes I feel most unattractive at the very moment people are telling me that they find me attractive."

By confronting Ginny with the discrepancy between the way that he and others experience her and the way that she experiences herself, Max helps Ginny open herself up to exploring her feelings about herself and how these feelings affect her interpersonal style. Here is another example.

Larry: "Dan, if I'm not mistaken, I believe one of the things you pride yourself on is being witty and humorous. I've often seen you as very funny, but sometimes I think that what you see as being witty I see as being cynical or even, perhaps, sarcastic. I'm not sure that you know when you go too far. It could be that I'm oversensitive about humor, and I think that it would be good to get feedback from some other people."

Dan: "You feel that sometimes I step over the line and maybe hurt others. I'd like to get a couple of examples, if I can. And I'd certainly like to hear what others have to say."

Dan hasn't experienced himself as cynical or sarcastic, but he's open to exploring the issue with Larry and with the other members of the group. Let's take one more example.

Vince: "Nancy, you see Cheryl as too pushy, making too many demands of you here. My feelings about Cheryl are a little different. I see her as gutsy. I think she makes demands of you, me, all of us. But what she asks is what's called for by the contract. Cheryl makes me uncomfortable at times, but I guess I need to be made uncomfortable."

Confrontation asks a person to consider another point of view. Obviously, the person doing the confronting believes that his or her point of view offers more promise or is more accurate than the point of view he or she is challenging. By being tentative, however, the confronter recognizes the fact that others might not share his or her point of view.

What are the sources of differences in experiencing? Discrepancies, distortions, evasions, game-playing tricks, and smoke screens, to name a few—and all of us are guilty of some of these at one time or another in our interpersonal lives. Let's look briefly at these sources of differences in experiencing.

a. *Discrepancies.* In all of us there are various discrepancies, for instance:

- between what we think and feel and what we say,
- between what we say and what we do,
- between how we see ourselves and how others see us,
- between what we say we're going to do and what we actually do,
- between what we are and what we wish to be,
- between what we really are and what we think ourselves to be,
- between what we communicate verbally and what we communicate nonverbally.

Here are examples of some of these discrepancies.

- You're confused and angry with me, but you say that you feel fine.
- You see yourself as playfully witty, but others see your wit as biting.
- You think that you're ugly, but actually your looks are above average.
- You say "yes" with your words, but your body language says "no."
- You say that you're interested in others, but you don't take the time to communicate basic understanding to them.
- You say that you want to improve your interpersonal style, but you don't put much effort into the group experience.

In this lab, of course, we're interested in confronting the discrepancies that affect interpersonal style and group participation.

214

b. *Distortions.* If we can't face reality as it is, then we tend to twist it so that it meets our own needs. How we see the world, then, is often more an indication of our needs than it is a true picture of what the world is like, including our interpersonal world. For instance:

- You're afraid of someone and therefore see him as distant, although in reality he's a warm, caring person.
- You have trouble relating to authority figures, and so you see the instructor or trainer as a god and write him or her off if he or she makes mistakes.
- Anne talks about herself in generalities, but you take this as adequate self-disclosure, because you're afraid of disclosing yourself.

Confrontation involves challenging such distortions. At its best, confrontation can help us break out of self-defeating views of ourselves, of others, and of interpersonal living.

c. *Evasions, game-playing tricks, and smoke screens.* If you're rewarded for game playing with other people—that is, if you get others to meet your needs by playing games with them—then you'll keep on playing. For instance, you might play the "Yes, but" game. In this game you ask others to help you solve some problems—for example, a problem in interpersonal relating. Once they begin to help you, you keep pointing out how their help really isn't helpful: *"Yes, but,* if I confront others, they'll turn away from me. I've seen it happen."* Another game is to seduce others in one way or another—whether sexual or not—and then get angry with them when they respond to your seductions. Game playing in interpersonal living is often a way of avoiding real intimacy. Real intimacy takes a great deal of work and is very demanding; games provide us with a kind of fake intimacy.

The best defense against game playing in your training group is to work toward genuine closeness with others. Effective group participants don't get hooked into playing games. For instance, if you refuse to play the role of helper in the group, you can't get hooked into the game of "Yes, but." When it does happen, game playing needs to be challenged.

Leon: "Margaret, I can trust you more than anyone else here. You listen to me, and I'm sure you care about me. I wonder whether you think that people here are gun-happy, too quick to confront."

215

Margaret: "I appreciate your trust, Leon, but I'm a little un-comfortable being singled out like this. I think that all of us should look at the issue of the amount of confrontation in the group."

Leon tries a game called "Pairing." He implies that he's OK and that Margaret's OK but that the others aren't. He tries to seduce Margaret into being his partner. This kind of pairing up could give him power in the group. However, Margaret refuses to go along, instead gently confronting Leon for wanting to play such a game.

3. *Strength confrontation.* Confrontation of strengths involves pointing out to a person the strengths or abilities or resources that he or she has but either fails to use or doesn't use fully. Here is an example of strength confrontation.

Vanessa: "Bob, when you talk to me, I really listen, because at times like that I feel that you're totally present. You attend; you listen; you understand very well. Your interactions with me are so rewarding that, when you fall silent for an extended period of time, I miss you. A certain lack of *quantity* of in-teractions on your part here seems to stand in the way of the *quality* of your presence. At least that's how I experience you. I'm just not sure why you seem to hold back so often."

Vanessa places a demand on Bob to make greater use of strengths he has used only partially. Studies have shown that strength confrontations are the most effective type. After all, there's something very positive about them, inasmuch as they deal with resources rather than with weaknesses. High-level communicators naturally and spontaneously use more strength confrontation than they do any other kind. If you are to use this kind of confrontation often, you need to school yourself to become aware of the strengths of others, which is the reason for the exercise on identifying strengths in others in Chapter 8.

4. *Weakness confrontation.* As the name implies, weakness confrontation consists of pointing out to someone what he or she either fails to do or does poorly. Although it's unrealistic to think that we can totally avoid this kind of confrontation in interpersonal living, still it's an ineffective communicator who uses this kind of confronta-tion the most.

Scott: "Sally, your silence here is beginning to drive me up the wall. It makes you seem so passive. When you're silent like that, I wonder what you're thinking. I begin to think that you're judging the rest of us. Your silence doesn't get *you* anywhere, either. It makes you seem disinterested, and then the rest of us just pick on you."

Since such confrontation dwells only on a weakness, it probably does little good by itself. The person being confronted usually either feels under attack and responds defensively or sits there and takes it all passively. Experience tells me that periodic weakness confrontation administered to a relatively silent group member does no good at all. With a little creativity on your part, you can make a weakness confrontation at least partially a strength confrontation.

> *Belinda:* "Sally, you've mentioned that part of the problem you have with silence is a fear of barging into conversations, interrupting others, and the like. Well, you can call it 'barging in' if you want to, but I'd welcome your interruptions. I see them as showing that you think enough about me to want to get in touch with me. And that's something I appreciate very much."

Whenever you're about to challenge someone, pause a moment and see whether you're emphasizing the person's strengths or weaknesses. If what you're emphasizing is almost exclusively a weakness, the odds are that your confrontation won't be very effective.

5. *Encouragement to action.* Encouragement to action involves urging someone to act in some reasonable way, to *do* something. High-level communicators are ordinarily *doers*. They initiate a lot in life, and they aren't afraid of having an impact on others or of letting others have an impact on them.

> *Marty:* "Ken, you and I have both admitted that we're much more passive than we'd like to be. I'm wondering whether we might enter into a little side contract with each other. Let's increase the number of times we initiate conversations with others here. We could help each other plan our agendas before each meeting and encourage each other to put our agendas into action during the meeting itself."

Not only does Marty confront Ken by encouraging him to act (there's a discrepancy between what Ken says he wants to do and what he's actually doing), but he suggests a way of going about it that

217

emphasizes *mutuality*. If Marty and Ken go on to help each other, then helping will take place without either of them becoming a helper or client.

Perhaps most of your confrontations will be mixtures of the kinds of confrontation I've just outlined. Whatever type of confrontation you use, the *way* in which you use it is crucial to its success.

The Manner of Confronting

Here are some practical hints that will help you make confrontation a more positive experience, both for you as the confronter and for the person you're confronting.

1. *Don't forget basic, accurate understanding.* The communication of understanding is the lubricant that makes interactions go more smoothly. Furthermore, the deepest confrontations are based on the deepest understandings. If you want to confront someone in any substantial way, you'll be successful to the degree that your confrontation is based on substantial understanding of that person. If, on the other hand, you don't try to understand a person before confronting him or her, your confrontation will probably be either ineffective or actually destructive. One excellent way to introduce a confrontation is to *begin* with a statement of basic understanding, as in the following example.

> *Ellen:* "Ed, you're just not used to disclosing yourself very intimately to anyone, and therefore even the kind of self-disclosure that seems to be easy for others here is hard for you. Talking about yourself makes you anxious. However, you and I shared an exercise once in which we identified our strengths. In our conversation you talked rather easily about what you saw as your interpersonal strengths. I wonder whether you might share some of that here in the bigger group."

Ellen's confrontation is a combination of strength confrontation and encouragement to action. However, she introduces it by communicating to Ed that she has some understanding of the difficulty he has in talking in the larger group.

2. *Be tentative.* I've already discussed tentativeness in connection with the communication of deeper understanding. Tentativeness shouldn't, however, be confused with *apologizing* for what you say.

218

Mike: "Gee, Adele, I don't mean to put you down, and I feel bad even having to tell you this, but . . . ."

Mike isn't being tentative; he's being apologetic. Such apologies show disrespect to the person being confronted, because they are actually ways of putting others down. The implication of the apology is "You can't take what I'm about to tell you."

Dorothy: "Jeff, you say that you swallow your anger a lot. *Could it be* that the anger you swallow *doesn't always* stay down? From the feedback you're getting, *it seems* that your anger does dribble out *somewhat* in cynical remarks or in silence or in your being uncooperative. *I'm wondering whether any of this rings true with you.* I know you can deal with strong emotion directly. I've seen you do it here."

This is a combination weakness/strength confrontation. The italicized words make it tentative but not apologetic. The word *balance* is an important cue here. Confrontations shouldn't be accusations, but neither should they be stated so tentatively that they lose their force.

3. *Know why you're confronting someone.* Ideally, confrontation is a way of getting more intimately involved with another person. If it's done well, it's a sign of caring and intimacy. However, if you confront somebody merely because you're frustrated or angry with him or her, you'll end up dumping your negative emotions on the person and, at least for the moment, destroying mutuality.

Teresa: "Ned, you're just rude and selfish. You say what you want whenever you want without consideration for anyone's feelings. I wonder why you're even here!"

This isn't confrontation; it's accusation and punishment. The point isn't that Teresa shouldn't be angry. But there are better and more constructive ways for her to show her anger. We'll see how a little later on in this chapter.

4. *Don't confront until you've earned the right.* Basically, you earn the right to challenge another by doing two things. First of all, you're open to confrontation yourself; you're the kind of person who is willing to put himself or herself on the line. Second, you make sure that you have a decent relationship with the person you want to confront. This means spending time trying to understand the person

and trying to communicate your understanding before challenging him or her. If you and the other person have a deep, caring relationship with each other, your confrontations can be quite strong without being harmful to your relationship.

5. *Don't gang up on a person.* If a number of other people have already confronted a particular person, you shouldn't add your two cents just because it seems to be the thing to do. Confrontations work a lot better if there's some space between them. Therefore, don't save up your confrontations and then cash them in all at once. Saving up confrontations means not being genuine and honest with others as you move through the group experience—or, for that matter, through life. If you don't confront at all, you'll be passive and perhaps dishonest. If you save up your confrontations, you'll probably end up being aggressive and hostile.

6. *Be concrete.* If your confrontation is vague, you can hardly expect the person you're confronting to do anything about it.

> *Anthony:* "You're too passive, Sam. You have to reach out for life if you expect to get anywhere."

Confrontations built upon your understanding of others will be as concrete as your understanding is.

> *You:* "Sam, when I talk to you, I pick up all sorts of nonverbal cues that you're with me, that you're listening to me. But I'd like you to reply to me verbally more often. I'd like to dialogue with you, and sometimes I catch myself in monologues with you. Maybe I contribute to the problem by just rambling on. How does this strike you?"

This is much more concrete than just telling Sam that he's passive. To put it briefly, all that was said about concreteness in Chapter 5 applies to confronting others also.

7. *Don't confront by using only nonverbal hints.* Nonverbal hints about what you're thinking and feeling aren't enough. If you fall silent, bite your fingernails, slouch in boredom, bite your lips, or look at the floor, you indeed send messages—some of which may well be confrontational—but these messages need to be backed up by words if they're to be direct and clear. Don't expect other people to read your mind.

8. *Confront only if you want to grow closer to the person you're confronting.* Confrontation is a sign not only that you have a relationship but that you want that relationship to grow. This doesn't mean

that you intend to become friends with your fellow group members outside the group. It does mean that you want to have a growing relationship with the other person inside the group. Both in the group and in everyday relationships, if you and another person feel that your relationship is stale, that it's not really growing or going anywhere, *try putting confrontation aside for the time being and finding out what's happening between the two of you.*

Describing behavior rather than punishing the other person. One hint is so important and practical that I'll explore it separately. Confront by *describing* a person's unproductive behavior and by *describing* its effects (what it does to the person, to you, to others). There's a strong tendency in most of us to substitute less useful and more punitive forms of verbal behavior for description. Some of these substitutes for description are:

- *commanding.* Confrontation is an invitation to explore behavior, not a command to change behavior.
  - *Commanding:* "Tony, don't keep asking Sandy how she feels. You're constantly doing that; it's like a cliché. Tell her how *you* feel and how she affects you. Then she might do the same."
  - *Describing:* "Tony, unless I'm mistaken, almost every time you ask Sandy how she feels—and you seem to do so fairly often—I see her make a face. You ask her to share her feelings with you, but I'm not so sure that I see you sharing your feelings with her. I even get the impression that you have a good idea how she's feeling before you even ask her."
- *judging, labeling, or name-calling.* Accusations ordinarily won't get you anywhere, whereas descriptions might.
  - *Judging:* "Peter, you're selfish! You get what you want here and leave the rest of us high and dry."
  - *Describing:* "You're very verbal, Peter, and very assertive. At almost every group meeting you begin with your own agenda. It's our fault, too, because we always wait for *you* to start. It seems that neither you nor the rest of us are doing anything to change this pattern."
  - *Accusing:* "Gene, you don't like Kay. You haven't given her a chance from the start."
  - *Describing:* "Gene, it's no secret that you and Kay have things to work out in your relationship, but I see her sitting there and taking things that would really tick me off. You don't initiate conversations with her. When she contacts you, your part of the

221

conversation is very brief. Even your voice strikes me as very disinterested. I'm in the dark. I don't know what's going on between the two of you."

- *questioning.*
  - *Questioning:* "Brenda, is it safe for you to reveal so much about yourself so quickly? Do you think that it might turn some people off?"
  - *Describing:* "Brenda, you've revealed yourself more than anyone else in the group. And now this evening you've begun to explore your sexuality—and it's only our third meeting. Frankly, it scares me when someone moves that fast!"
- *being sarcastic or cynical.*
  - *Sarcasm:* "Boy, Paul, you really know how to put people at ease!"
  - *Describing.* "Paul, now that Craig has stormed out of here, I wonder whether you have some idea what angered him so. Briefly, here's my view of what happened. You told him quite frankly that he's not the kind of person you usually make friends with. But you did it in such a cool voice. Neither your voice nor your body showed much, if any, feeling. I shuddered a bit inside and began wondering what was happening inside Craig. Your expression hasn't changed since he left, and I'm not sure what's happening inside you."

Descriptions of behavior have a kind of objectivity about them that helps the person being confronted to avoid the kind of natural defensiveness that arises with confrontation.

   The "MUM" effect. Some people seem to be too ready to confront others, and many of these people seem to confront quite poorly. Others don't want to engage in confrontation at all because of the emotional price they would have to pay. Part of the emotional price of high-level confrontation is that it involves a degree of intimacy that some people find too demanding. Another part of the emotional price is the feeling we experience when we have to give others some bad news. In a sense, confronting others does involve telling them some bad news. Even strength confrontation involves telling the other person that he or she *isn't using* a talent or ability or resource that he or she has. Most of us are very reluctant to be the bearers of bad news. Psychologists call this reluctance the "MUM effect." In ancient times, messengers sometimes were killed when they brought bad news to the king (although this doesn't seem to have been a very efficient way of getting rid of the bad news). As you can imagine, messengers became more and more reluctant to deliver bad news. Some of that reluctance, it seems, has passed into our bones. True, we're not killed when we give others a piece of bad news, but we feel very uncomfortable all the same. It's not just that we don't want to make the other person feel bad; it's also that *we* feel bad when we make another person suffer, and we avoid it if we can. Confrontation is also one of the strongest ways in which we influence one another. If you're uncomfortable about having some influence in the life of another person, you'll likely be uncomfortable with confrontation. Since "nice guys" don't make others suffer and don't make demands on them, "nice guys" avoid confrontation.

In the best relationships, confrontation is not a specialty. It flows naturally from the relationship of two people who are committed to each other. If you see yourself falling victim to the MUM effect, you can ask yourself whether you're really committed to those whom you refuse to confront. You can ask yourself whether you've worked at establishing the kind of intimate relationship that enables you to confront reasonably and responsibly. In the best relationships, people move from understanding to challenging so easily that they would be surprised if you told them that they were engaging in "effective confrontation."

Responding Growthfully to Confrontation

Even when it comes from someone who has our interests at heart, confrontation can make us defensive. No one says that it's easy to be challenged. Now that we've discussed how to confront *others* in a responsible way, it's time we asked "How can I *receive* confrontation in a way that will help me grow as a person?"

Defensive responses to confrontation. Since confrontation can make you uncomfortable or make you feel inadequate, you can find ways to avoid responding to the confrontation itself while getting rid of the uncomfortable feelings. Here are some of the defensive tactics people use to avoid a confrontation.

- *Refusing to think about it.* For example, you can listen to a confrontation, say "I see," and then proceed to put it out of your mind. This seems to happen with some frequency in human-relations-training groups.
- *Hearing what you want to hear.* Someone tells you that you're an attractive person and that you have good verbal

skills that you don't use enough. You hear the part about not using your talents, but the rest seems to slip your mind. You forget the positive and get depressed by the negative and do nothing.

- *Letting the confronter know that he or she isn't perfect either.* The strategy here is this: if you discredit the person confronting you, you don't have to listen to or respond to the confrontation. Who wants to listen to people who have problems of their own?

- *Getting the confronter to change his or her mind.* One way to deal with the confronter is to win him or her over—to show that you're really not a bad person, even though you do have your bad moments, and so forth—gently persuading the confronter to see your side of things.

    Greg (having been confronted): "I know I don't express much feeling here, Judy, but I'm not sure that it's called for—not as much as some people seem to ask for, at any rate. After all, there *is* something artificial about these groups. I don't want to manufacture emotion. In fact, that's part of the contract. I want it to flow naturally."

    The clever person can "reason" his or her way out of any confrontation.

- *Merely rejecting the confrontation.* Here you tell the person confronting you that he or she is wrong and let it drop at that. This isn't a very subtle way of getting out of being confronted, but it can work, unfortunately.

- *Getting the other members of the group on your side.* You can probably always find people in the group who will take your side when you're being confronted.

    Chuck: "Do you all see me as sarcastic? Ellen, I can't even imagine us being sarcastic with each other."

- *Agreeing with the confronter.* You can immediately agree with what the confronter has to say. It usually gets the confronter off your back, and your "honesty" wins the approval of the other group members. If you agree with the confronter (meanwhile doing nothing about what he or she is confronting you on), it's difficult for the person to confront you again. After all, you've already admitted that what he or she has to say is true.

It's possible to use a combination of these strategies for getting out of confrontations and leave the group without ever having dealt honestly with any confrontation. But the supposition of this book is that anyone who did so would come out a loser.

Creative response to confrontation. If someone overcomes the MUM effect and takes the risk of confronting you reasonably and responsibly, mutuality requires that you respond openly and directly. Here are some hints to help you respond constructively to confrontation.

1. *First make sure that you understand what the other person is saying.* An excellent initial response to a confrontation is *basic, accurate understanding.* Respect for the other person (in this case, the confronter) requires that you get his or her messages straight.

> *Donna:* "Barbara, when someone here disagrees with you, you usually have a lively conversation back and forth for a minute or two. But I've noticed that then you usually fall silent. Sometimes you don't say anything for the rest of the group meeting. I think I see this as a pattern with you. When you're silent, I don't know what's going on inside. And up until now I haven't tried to find out."

> *Barbara:* "You see me retreating, as it were, after a brief contest. And you're not sure whether I'm angry or licking my wounds or what. And I guess you're not comfortable with my just retiring like that."

Here Barbara merely makes sure that she's getting the right message. Because of the emotions involved, it's easy to distort and misunderstand the message. So often we hear what we want to hear. We hear invitation as command. We hear suggestions as attacks. Thus, even reasonable confrontation can *feel* like an attack. If you use basic understanding as an initial way of responding to confrontation, it will give you time to reflect and perhaps to let some of your initial defensiveness settle.

2. *If you're confused, clarify the confrontation.* If the confrontation upsets you, it's best to say so and to deal briefly with your feelings. This kind of response is open rather than defensive. It also gives you a chance to catch your breath so that you can respond reasonably. If you don't understand the confrontation, ask the confronter to explain and to use examples. Get the other members of the group to help you see the point. Sometimes a different member will be able to make the point more clearly than the original confronter.

225

3. *Explore the confrontation nondefensively with the confronter and with the other members of the group.* Once you understand what's being said, decide whether this is the way you see yourself, and find out from the other members of the group whether they see you the same way as the confronter does. The point of this checking is not to get others on your side but to see whether what the confronter says is generally a problem. For example, if the confronter suggests that you initiate conversations so seldom that lack of quantity seriously affects the quality of your participation in the group, explore how you feel about your level of initiation, and find out how others see you and how they feel about your interactions.

4. *If possible, experiment with new behavior if that seems reasonable.* As I said in the first chapter, human-relations-training groups are places where you can experiment with new ways of behaving interpersonally. For example, you might gradually increase the number of times you initiate conversations with others to see how you feel about it. You may like this new, more assertive you. If so, you're a winner. Or you may find out that you don't want to be that assertive, which could be an equally important discovery. The ultimate choice is yours, of course, but experimenting reasonably with new interpersonal behaviors will help you make your choice.

5. *Don't try to take care of the confrontation all at once.* There's a difference between ignoring a confrontation and giving yourself a reasonable amount of time to let it sink in and affect your behavior. Very often an initial confrontation will get an issue out on the table; in subsequent meetings the issue can be explored more fully, and you can begin to experiment with your behavior.

Self-Confrontation. Up to this point, I've emphasized challenges among group members. However, mutuality will be promoted if you and your fellow group members also learn how to challenge yourselves. If you confront yourself in the group, you'll be less defensive with the issues that others bring up. You won't force others into the position of constantly challenging you. And you'll put the responsibility for change directly where it belongs—on yourself.

> *You:* "I notice that, whenever anyone expresses affection for me here, I pull back. At least I feel myself pulling back inside. I don't know whether others notice this—though I imagine that it comes out, at least in my nonverbal behavior. I think it's about time that I talked about my fear of affection—of both giving and receiving it."

226

If you challenge yourself, you can then call on the resources of the group to help you explore, monitor, and change your behavior. You can call for support, feedback, and further challenge. Ideally, the amount of self-confrontation in the group becomes greater than the amount of confrontation by others. Such a development is a sign that people are taking responsibility for themselves. Self-confrontation also helps make the group a group of equals instead of a group of helpers and clients.

Exercises in Confrontation

Exercise 37
Self-Confrontation, Good and Bad

Before confronting others, you can practice on yourself. The purpose of this exercise is to give you a feeling for both responsible and irresponsible confrontation.

Directions

Think of a few areas in your interpersonal life in which you could benefit from some kind of challenge or confrontation. Next write out a statement in which you confront yourself *irresponsibly*. In other words, deliberately break the rules of reasonable confrontation outlined in this chapter. Then write out a statement in which you confront yourself *responsibly*, observing the hints I've given.

Example

a. *Irresponsible self-confrontation:* "You're so parental! You know that you're too controlling, and yet you do nothing about it. Your behavior is arrogant and insulting. I don't think that you have a basic respect for people. Sure, you've been called an attractive guy, but at the core of you there's a great deal of selfishness and inconsiderateness."

What makes this confrontation irresponsible?

b. *Responsible self-confrontation:* "There are two or three behaviors—apparently characteristic of your interpersonal style—that you might want to reexamine. First of

227

all, you talk a lot about how we should all be observing the contract. Usually your contract talk doesn't involve any dialogue with anyone, with the result that it sounds like a sermon at times. Second, you tend to praise people a lot. Even though you use positive labels like 'understanding' and 'good group member,' you're still labeling. Often the tone I find in your voice is one of a parent speaking, if not to a child then perhaps to an adolescent. I wonder whether you're in touch with these or any other behaviors that you might see as parental."

Now write out four pairs of self-confrontational statements similar to those in the example. Choose areas of your life that are related to your interpersonal style, areas that you'd like to work on. Each time you write an irresponsible statement, indicate afterwards the things that make it a poor confrontation.

Share these statements with your fellow group members in the way determined by your instructor. Check to see whether your "responsible" confrontations are seen as responsible by others. What words or phrases did you use that indicate tentativeness? Is the confrontation concrete enough?

Exercise 38
Changes from First Impressions

Directions

Try to think of the first impressions you had of each of the other members of your group. Write these down. Next write down ways in which your first impressions have changed, if indeed they have. For each person, write down something positive and something negative. In each case, choose something you think will help the member pursue the goals of the training group.

Example

"Leo, my first impression of you was that you were a shy, rather passive, but 'nice' person. You seemed to hesitate to speak up, to initiate conversations with others. I noticed early in the lab that you blushed once or twice when people asked you to be more assertive. This confirmed my impression of

228

you as being shy. However, as the group moved on, what I thought would be a weakness proved to be a strength. You don't respond to others too quickly, but I see your taking your time as showing a deep respect for others. When you do interact with others, it's with intensity, conviction, and care. I sit up and listen when you talk to me. You're not showy, but you're not passive either. I've seen you get angry, control your anger, and then let it out responsibly—that is, you've used your anger in a way that has helped you to get closer to others. However, although I don't see you as passive any more, you're still a bit too deliberate for me. You seem to be almost too careful."

Do one of these comments in writing for each of your fellow group members. You don't have to give a summary of the entire personality and behavior patterns of each. Rather, choose one area that you think may have special meaning for the other. You might use the following formula:

"In the beginning I saw you as _____, but now I see you as _____, and the following specific things have caused me to change my impression of you: _____.

Or the following formula might be more appropriate in some cases:

"In the beginning I saw you as _____, and this impression has been confirmed for me by the following experiences with you: _____."

This exercise should not consist of feedback or confrontation just for the sake of confrontation. Rather, try to make your statements useful by helping the other person explore his or her interpersonal style and by providing background for exploring your present relationship with that person.

Exercise 39
Self-Confrontation: The Quality of Your
Participation in the Group

As I've said, learning self-confrontation is a part of this group experience that helps greatly in making it a mutual experience.

229

Directions

Examine the quality of your participation in the group experience so far. Write out five brief self-confrontational statements, one in each of the following categories:

1. information confrontation,
2. experiential confrontation,
3. strength confrontation,
4. weakness confrontation, and
5. encouragement to action.

All of these statements should relate to the quality of your participation in the training group.

Example

- *Information:* "I don't fully understand the function of the instructor or trainer in these groups. I'm especially confused about whether he or she should be putting himself or herself on the line in the same way as the rest of us. Because I haven't cleared this up, I've hesitated to interact with our instructor. I even resent her a bit."
- *Experiential:* "Sometimes, when I'm bored, I assume an attending position in the group—not to get involved but to avoid being called for lack of involvement. I experience myself inside differently from the way I'm acting."
- *Strength:* "I have the ability to talk easily with the more silent or nonparticipating members of the group. Yet I seldom use this ability. I see a lot of potential in a number of the people who don't participate much, but I don't do very much to help them get at it. I've been lazy."
- *Weakness:* "Henry still frightens me in a number of ways. He's a very strong, sometimes aggressive person. With him I'm merely passive. I just stay clear of him."
- *Encouragement to action:* "I want to reveal myself more deeply. I've gotten very comfortable in here. Therefore, the group isn't very intense for me any more. I'd like to talk about my fears—my fears of being a trainer, my fears of getting too close to other men, my fears of being rejected. Talking about these will make this experience much more real and much more intense for me."

Share what you write in the way determined by your instructor. This is an exercise in self-awareness, and it will provide

230

you with material for self-confrontation. You can use this exercise repeatedly to give yourself self-confrontational material.

Exercise 40
Confrontation of Strengths

As we've seen, high-level communicators confront other people's strengths much more frequently than they do their weaknesses.

Directions

In your mind's eye, see each of the other members of your group working at his or her best in the group. Write down what you see for each person. Obviously what you "see" is in part what you hope for each member, but presumably your hopes are based on the potential, abilities, resources, and talents of each person, even though he or she might not be using them fully as yet. Be as concrete as possible.

Example

"Even though Rich started out on the wrong foot with both Paula and Bruce, I see him as being capable of mending fences and of developing good relationships with both of these people. He got on their bad side by challenging them without first understanding them. Even though they have more or less rejected him and don't talk to him much, he has remained open without becoming bitter. He joins in conversations they have with others, and he has the guts to talk to both of them directly. He has corrected the mistake of challenging without first communicating understanding to others. A certain good-hearted simplicity got him in trouble with Paula and Bruce, but I think that this same simplicity will be a resource to help him mend fences."

In this exercise it isn't necessary to spell out every unused strength the other person has. Choose one or two strengths, talents, or abilities that you think, if put to better use, would make a difference in this person's participation in the group and in his or her interpersonal life. Share your strength confrontations in the way determined by your instructor.

231

As you yourself receive feedback, see whether the others tend to agree on which strengths you don't use fully. Do you think that you have the strengths that others see in you? How do you want to develop these resources?

Exercise 41
Confrontation: Safe and Anxiety-Producing Issues

Not all self-disclosures carry the same weight. Each of us has some particularly sensitive areas we don't like to talk about. We get anxious when we do talk about them. This exercise is designed to help you and the other members of your group get in touch with these more sensitive areas and to help you talk about them in the group *if* they would be *appropriate* self-disclosure. (This exercise is not intended to force you to disclose your secrets.)

Directions

Write down two lists for yourself and for each of the other members of your group. In the first list, put safe topics—things you can talk about easily in the group. In the second list, put anxiety-producing topics, topics that make you anxious when you talk about them.

Example

- *Myself—safe topics:*
  - how I feel about myself
  - my relationship with everyone in the group except Bill
  - my fears about being assertive
  - my interpersonal laziness when I'm with people I feel comfortable with
- *Myself—anxiety topics:*
  - my relationship to Bill
  - my sexual feelings about myself and in the group
  - my tendency to become dependent on others
  - my fears about not being attractive
- *Group Member A—safe topics:*
  - her interpersonal values, if we don't get too specific
  - her relationship to all the members of the group
  - her fears about being lonely; her fears about being too aggressive

- her friendships outside the group and what she learns about herself from them
- *Group Member A—anxiety topics:*
  - how she feels about herself
  - her defensiveness
  - her cynicism and sarcasm
  - how she feels about the instructor and other authority figures

When you draw up your list of safe and anxiety-producing topics for others, you'll often be operating on hunches. Remember to be tentative when you communicate these hunches to others. Also, choose topics that affect the person's interpersonal life and his or her participation in the training group.

The Skill of Immediacy: "You-Me" Talk

Immediacy, or "you-me" talk, is one of the most important but also one of the most difficult interpersonal skills. This chapter discusses the two kinds of immediacy: relationship immediacy ("Let's talk about how we've been relating to each other recently") and here-and-now immediacy ("Let's talk about what's going on between you and me right now as we're talking to each other").

Relationship Immediacy and Here-and-Now Immediacy

Relationship immediacy. "Relationship immediacy" refers to the ability to discuss with another person, directly and mutually, where you stand at the present in your relationship to him or her and where you see the other person standing in his or her relationship to you. Since one of your goals in the lab group is to establish and develop relationships with the other group members, learning the skill of relationship immediacy is very important.

> *Dwayne:* "Jack, you and I haven't talked directly with each other for two or three sessions. I notice that we—or I should say I—have been avoiding even nonverbal communication. I feel that something has gone wrong between us but that we're not admitting it or looking at it. A few weeks ago we got into an argument about my being passive and your being aggressive. We didn't fight very fair with each other. But I think we still respect and even like each other, even though we've stopped communicating. I'm dissatisfied with the way we've been relating—or not relating—for the last few weeks. I'd like to sort things out with you."

In relationship immediacy, people talk about where they stand with each other generally, not just about what's happening at the present moment.

Here-and-now immediacy. "Here-and-now immediacy" refers to the ability to discuss with another person what's happening right now in *this* present conversation. Here is an example of this kind of immediacy being used in the context of a husband/wife relationship.

Eileen: "Tom, I feel that you're edgy. I guess I feel that we're both edgy with each other, even though neither of us is saying anything about it. Every time we talk about how we differ in disciplining the kids, we get uptight. I don't want to get angry at you and then just run off. I wonder what we can do right now to talk about the problem of how we differ in discipline, and what we can do to talk about it reasonably."

Eileen isn't talking about the whole sweep of her relationship with Tom; rather, she's concerned about what's happening in the here and now of *this* conversation.

Immediacy: A Complex and Difficult Skill

Being good at both relationship and here-and-now immediacy is an important part of being socially intelligent. Like other human-relations skills, immediacy has three parts: (1) an awareness part, (2) a communication-know-how part, and (3) an assertiveness or "guts" part.

The awareness part of immediacy. If you're going to talk to someone about what's happening between the two of you—either in your overall relationship or in a here-and-now conversation—you have to *know* something about what's happening. You have to have some idea of what's taking place between the two of you, and you have to be aware of how it's affecting you. For instance, if you don't pick up the nonverbal and voice-related cues that another person is angry with you, you can hardly bring up the issue of his or her anger. Therefore, the skill of immediacy calls for the kinds of awareness that are involved in both *basic understanding* and *deeper understanding*. This is one of the reasons why immediacy is a complicated skill: it depends on your being good in all the other skills we've discussed to this point. If you find yourself saying such things as "I didn't even know that she was angry," then you have work to do on your attending behavior. It takes a great deal of interpersonal sensitivity to stitch together the kinds of awareness that prepare the way for immediacy. Let's take an example.

- You see that Connie isn't looking at you when she talks to you.
- You notice that you've been on edge since Connie came into the room.
- You remember that the last time the two of you talked you were sarcastic with each other. You had a "good time" making fun of each other, but now you begin to wonder whether it was so much fun after all.
- You notice that you're both reluctant to talk directly about your uneasiness with each other.
- You notice that you feel bad because there's some kind of strain between the two of you.

Once you're aware of all these things, you can begin to do something about them.

The communication-know-how part of immediacy. The skill of communicating immediacy is formed from a combination of four other skills: basic understanding, deeper understanding, self-disclosure, and confrontation.

• *Basic understanding and deeper understanding.* The communication of the kinds of awareness I've just discussed involves the communication of both kinds of understanding.

Clara: "I don't think I want to go to the movies tonight. I've had a slight headache all day. The kids in school were devils today. And we had another one of those useless faculty meetings. I don't know why we don't rise up in revolt against the principal."

Kevin: "You feel too drained to enjoy a movie. I wonder whether the headache comes just from teaching those kids. You and I had an argument a couple of days ago that we never talked through. I have a sneaking suspicion that we're both still under a strain because of that argument."

Kevin responds at first with basic understanding, but then he moves deeper. He bases his deeper understanding on the wider context of his relationship with Clara. He puts together a few cues and offers a hunch. Communicating deeper understanding doesn't mean becoming a detective. It does mean noticing what's happening in your relationships with others.

• *Self-disclosure.* Communicating immediacy also involves revealing how you think and feel about what's happening in your relationship with the other person. Disclosing yourself is extremely important; otherwise you tend to treat the other person as someone who needs help, suggesting that it's his or her problem and not one that involves both of you. To continue with our example, Kevin goes on to say:

> "Anyway, I guess I'm saying that I feel that you and I still have some unfinished business. I was pretty nasty in our argument the other night. I think I feel a little ashamed, and that's why I'm reluctant to talk about it."

Immediacy isn't a way of handling or "dealing with" the other person; it's an exercise in *mutuality.* Your self-disclosure, of course, should be appropriate; that is, it should be related to what's happening between you and the other person.

• *Confrontation.* Immediacy is always an *invitation* to talk about a relationship, and therefore it has a challenging or confrontational quality about it. In our example, Kevin continues:

> "I think that maybe both of us would like to clear the air. I'd like to know whether you feel any of this too, and I'd like some help in getting my own feelings straightened out."

Kevin confronts Clara in the spirit of mutuality by an encouragement to action—in this case, the action of talking things out with him.

If you're poor at any of these four fundamental skills, you'll be poor at immediacy. But without immediacy people tend to engage in interpersonal game playing with one another. Game playing is a low-level communicator's substitute for immediacy and intimacy.

The assertiveness part of immediacy. As with other interpersonal skills, immediacy isn't just a matter of good awareness coupled with good communication know-how. Perhaps more than any of the other interpersonal-communication skills, immediacy demands guts. Many of us wait too long to be open in a relationship. For instance, various annoyances crop up between a husband and wife. They vaguely notice the uneasiness in the air but don't say anything about it. Communication grows stale or edgy, yet the two of them still say nothing about it. Soon minor annoyances become major, and husband and wife begin to question each other's motives. Bad feelings grow until it all bursts forth in a big fight—what someone has called the game of "uproar." The couple fight; emotions are spilled out in hurtful ways. Finally the anger is spent, and things settle down. But nothing has been learned, and they go on their way until the next game of "uproar."

The assertiveness or courage necessary to face a small interpersonal problem in the beginning, when things can still be managed, just seems worth the effort. As we've seen, games substitute for intimacy. In the case just described, real intimacy calls for the use of immediacy early in the relationship.

Since building relationships is one of the goals of your training group, it goes without saying that immediacy is an important communication skill within the group experience. However, perhaps because immediacy is such a demanding skill, there's a tendency to use it infrequently, both inside the group and outside. As a result, even toward the end of a human-relations-training group, *the members are still unsure of their relationships with one another.* "Where do you and I stand with each other in our relationship?" is a question that is too infrequently asked.

Immediacy and the Interpersonal Gap

Immediacy is a skill that helps us bridge what one psychologist (Wallen, 1973) has called the *interpersonal gap.* Let's explain this term and give an example.

The Communication Process

1. *What you want to accomplish.* When you speak to others, you usually intend to accomplish something. A person can't know what it is you want to accomplish unless you tell him or her, either through words or through your behavior.

An Example

You're listening to Joan talking about what a hard time she's having in the group. You want to encourage and support her. This is what *you want to accomplish* by speaking to her. You want to let her know that you're "with" her.

239

The Communication Process

2. *Putting your message into words and/or nonverbal behavior.* Since you accomplish nothing unless you communicate, your next job is to translate your intention into words or nonverbal behavior. You expect your words or your behavior to accomplish what you want to accomplish.

3. *The other person receives and interprets your words and your behavior.* The other person listens and observes your behavior. The problem is that this person may see you as trying to accomplish something *different* from what you think you're trying to accomplish.

> *This is the point of the interpersonal gap.*

4. *The effect your words and your behavior have on the person to whom you're speaking.* When you speak, you hope that your words will have the effect you intend. If they don't, you can be surprised, confused, and frustrated.

An Example

You reach out and touch Joan on the arm. In words, you communicate as best as you can basic, accurate understanding of her feelings, of the hard time she's having in the group. By touching her and by communicating understanding, you expect to communicate to Joan that you're "for" her and "with" her.

Joan takes your communication of basic understanding in the wrong way. She sees your touch and your words as a way of being parental. She thinks you're acting in a condescending way. *She misunderstands what you're trying to do.*

Joan, instead of feeling understood and encouraged, feels put down and inferior. She is annoyed with you and resentful. She wishes you'd keep your remarks to yourself. She might just sit there feeling all of these things, or she might tell you how she feels.

If you feel that you've been misunderstood, if you feel that some kind of interpersonal or communication gap has opened up between you and another person, then the skill of immediacy can help you bridge that gap.

> *Joan:* "I really resent what you've just said. It's so damned parental."

240

You: "Something has gone wrong here, Joan. I had no intention of being parental. I hate it when people are parental to me. I'm really thrown. I'd like to find out how I got my message messed up."

This kind of remark is here-and-now immediacy, designed to take a look at what's happening *right now* between you and Joan. If you want to bridge the interpersonal gap, each of you must see how *the other* sees your interaction. This is precisely what you're trying to do here through immediacy. If Joan doesn't respond to you, you can call on the resources of the group. The other members can help you and Joan find out why you're missing each other.

Immediacy and the Interpersonal Loop

If I speak to you and then you speak to me, we've made an interpersonal or communication "loop." Let's look at an example of what can go wrong in the course of this loop.

The loop begins: I speak.

- *My intention:* I want to let you know that I care about you.
- *My words and behavior:* I kid around with you. I'm playful with you—or so I think. I mock you and cut you up playfully. I'm not playful with people unless I like them, but you may not know that.
- *The way you read my message:* You see me as just being silly, goofing around in the group when we're supposed to be serious with each other.
- *The impact I have on you:* You're annoyed. You feel that I'm running away from the business of the group, which is establishing deeper relationships. You don't want to kid around with me.

The loop closes: You respond.

- *Your intention:* You want to let me know that my behavior is inappropriate. You think I'm good-hearted, and you don't want to hurt me.

241

- *Your words:* You confront me as carefully as you can, say-ing "I usually enjoy your humor a lot, but I'm not sure that I like it in the group. I like being serious with you, getting close to you. So your fooling around puts me off."
- *The way I read your message:* I feel that you've ignored my real message, which is that I like you. I see your confronta-tion as a way of putting me off.
- *The impact you have on me:* I'm annoyed and hurt. I feel distant from you now.
- *My response to you:* I mutter something about being sorry and say that it won't happen again. I fall silent.

Here the loop ends.

How could immediacy-type statements have made this loop more productive? There are a number of ways in which both you and I could have done things differently.

- *My words:* I could have let you know more immediately and directly that I care about you:
  Me: "I feel awkward. I want to let you know that I care about you, but I don't know how, except bluntly like this. I don't want my bluntness to turn you off."
Here my words carry the "you-me" message directly.
- *The way you read my message:* You could have looked at my fooling around from two points of view: (1) you could have seen it as inappropriate, but (2) you might also have won-dered what I was really trying to say by my fooling around. Another way of saying this is that you could have looked at my fooling around with *deeper understanding.* You might not have been able to see that it really meant "I like you," but you might have had a hunch that it was more than mere fooling around.
- *Your words:* You might then have responded differently to me.
  You: "My first impulse is to get annoyed at you when I see you fooling around during group time. But I guess I don't know what you may be trying to say by kidding around with me."
In other words, you could have admitted your own confusion without accusing or blaming me.
- *My response to you:* Instead of withdrawing from the in-

242

teraction with hurt feelings, I then might have said something like:

"I feel embarrassed. I fool around only with people I like. I didn't mean to goof off, but I have a hard time letting people know that I like them."

Immediacy-type responses would thus have made this interpersonal loop smoother and more direct.

As with confrontation, a high-level communicator uses immediacy instead of:

- *name-calling.*
  not: "You're aloof. You're in a world by yourself!"
  but: "You've stopped calling me up. I feel you're pulling away from me. It hurts me."
- *expressions of approval.*
  not: "You're wonderful!"
  but: "When you're not sure what I'm feeling, you check it out with me. I like that. I can trust you."
- *accusations.*
  not: "You never listen to me or to anyone."
  but: "You just missed what I was trying to say. That annoys me."

Cautions. Immediacy doesn't mean always looking for hidden meanings or always putting relationships under a microscope. Like other interpersonal skills, immediacy becomes a weapon (or merely boring) in the hands of a low-level communicator. Low-level communicators sometimes manage to execute the communication know-how part of a skill fairly well, but in the process they lose the person with whom they're trying to communicate. Communication know-how doesn't take the place of awareness or of social intelligence.

In focusing on immediacy as a separate skill, as I am in this chapter and in the exercises that follow, I'm taking it out of context. Immediacy might then seem to be something that's added on to interpersonal conversations. A high-level communicator, however, integrates immediacy into his or her relationships in an ongoing way. Immediacy, then, isn't constantly a "heavy" issue. A high-level communicator works at his or her relationships without becoming preoccupied with them.

Exercises in Immediacy

These exercises assume a relatively high degree of motivation to become involved with others. They aren't magical substitutes for the work that must go into building relationships.

Exercise 42
Similarities and Differences

This exercise is designed to help you sharpen the kind of awareness that is needed for immediacy.

Directions

Think of each member of your group, one at a time. Write down a few ways in which you're like each person in terms of your interpersonal style and a few ways in which you think you differ from each. Be as concrete as possible.

Example

Ways in which I'm like John

- We're both overly quiet in the group.
- We're both soft-spoken.
- When we do respond, it's usually in terms of basic understanding rather than in terms of confrontation or immediacy.

Ways in which I differ from John

- He uses more self-disclosure in the group than I do.
- I have a quick temper, which I usually keep under control, whereas John takes a long time to get angry.
- I fear the instructor and manifest it by resisting the exercises, whereas John also fears the instructor but manifests it by being passive and doing what he's told to do.

Next, briefly write down how you think these similarities and differences affect your relating to each other.

Example

- Because of our similarities, John and I haven't gotten to know each other very well. We don't initiate conversations with each other very easily, and we never challenge each other.

- Because of our differences, my feeling is that John doesn't care much for me. My guess is that he doesn't like my self-centeredness, my hidden temper, or my resistance. Sometimes I think he's too "nice," and he makes me feel guilty.

Finally, share your reflections with each of your fellow group members in the way determined by your instructors.

Exercise 43
Some Basic Immediacy Questions

Directions

When you want to review a relationship with anyone in your group, read the following questions, and check those that seem important for you to answer in order to determine where you stand with that person. Answering these questions should provide you with some concrete issues for discussion.

Immediacy Questions

1. How frequently do I initiate conversations with you?
2. How would I describe my investment in you?
3. How do I think you're invested in me?
4. What am I willing to tell you about myself?
5. What do I feel I can't tell you about myself?
6. In what concrete ways do I trust you? How do I feel when I entrust myself to you?
7. Do I work at responding with basic, accurate understanding when you reveal yourself to me?
8. Are there ways in which I feel phony when I'm relating to you?
9. If I challenge you, how do I do it? In what areas am I afraid to challenge you?
10. Do I know in what ways you'd like to change? In what ways could I be a resource for you in your desire to change?
11. Do I give you straight feedback on your interpersonal style?

12. Do I initiate immediacy ("you-me") issues with you? What immediacy issues am I afraid to bring up? What issues do both of us avoid?
13. How intense is our relationship? How intense would we like it to be? Do we have different wants? How do we handle our differences?
14. What does our relationship lack?
15. What could we do to improve our relationship? Do we want to invest the time and energy it would take to improve our relationship?
16. If we have a good relationship, what are we doing to maintain its quality?
17. How do our other relationships affect our relationship?

What other questions should you ask yourself to find out where you stand with the other person?

Exercise 44
Summarizing as Preparation for Immediacy

In order to determine where you stand with another person (relationship immediacy), it's useful to review the highlights of the development of your relationship.

Directions

Picture each member of your group in your mind's eye, and write out a short summary of the highlights in the development of your relationship. Write your summary as though you were talking directly to the other person.

Example

"Norm, at first you seemed too colorless for my tastes. You were quiet and, I thought, timid. I tended to ignore you, even when you did speak. I wasn't giving you a chance. In my stupidity, I didn't think that it was necessary to relate to you, because I didn't think that you were going to be an important person in the group. All of that obviously says a lot about me. But, as I began to get off my high horse and tried to relate to others as a member of a community, I began watching you as you interacted with others. You were 'quiet' in some sense of the word—mild-mannered, maybe—but hardly weak or timid. You listened well to others and responded with understanding and feeling. It suddenly dawned on me that you were a much more effective member than I was. I got down on myself, and I still feel phony when I compare myself to you. Recently I've wanted to talk to you more directly, but I'm ashamed to do so because of the way I reacted to you in the beginning. You don't tend to initiate conversations with me. I'm not sure why. Maybe you think I'm not interested. Obviously you no longer appear colorless to me, but I'm uneasy with you. I don't know how to relate to you."

When you share your summary with another, see what similarities and differences there are in the ways the two of you experience your relationship. What do you want to change in each relationship?

The Skills of Being
an Effective
Group Communicator

High-level communication in groups is complicated and demanding. This chapter reviews the kinds of skills necessary for engaging in this demanding work. These skills are called group-specific skills, because they are more than just the skills you use in your one-to-one conversations with others.

The three sets of skills I've discussed and illustrated so far—the skills of self-presentation, the skills of responding, and the skills of challenging—are the building blocks of good group participation. If you're good at these skills, you have the basic resources you need to be an effective group communicator. Being good at them doesn't mean, however, that you'll *automatically* be an effective communicator *in a group*. To be an active, effective communicator in a group, you also need a special set of skills, called *group-specific skills*. Perhaps *the* group-specific skill is putting all the skills you've learned up until now into practice in free-flowing conversation in a group. So first I'll review the skills you've just learned *as they relate to your group participation*. Then I'll discuss additional group-specific skills.

Using Self-Presentation Skills in a Group Setting

Facing the issue of trust in the group. Most people find it more difficult to talk about themselves in a group setting than in a one-to-one conversation. Perhaps you've had very little experience talking about yourself personally in groups. If so, the idea of disclosing yourself to a number of people at the same time may take some getting used to. Moreover, you may trust some members of the group more than others. If this is the case, you might limit your self-disclosure, tailoring what you say to the person you trust the least. What can you do if you trust some people in the group more than others? First of all, you can use the skill of immediacy. If you feel that you can't trust someone as much as you'd like to, you can use "you-me" talk to work out the issue of trust between you and the other person.

> *You:* "Roberta, a few people have told me that I'm still a hidden person in many ways. I just don't talk about myself in the group as intimately as other people do. I think one reason I don't is that I've never worked out some feelings of suspicion I have about you. Once, early in the group, you confronted me pretty harshly. Even if you didn't mean it, it hurt, and I think I'm still licking my wounds. You backed off, too; so we don't talk to each other much. I have to clear this up if I'm going to feel free to talk about myself very intimately."

If lack of trust is a group-wide issue, it should be faced directly by all the group members. One indication that the level of trust isn't what it might be in a group is that the *intensity* of self-disclosure hits a certain safe level and then stays there.

> *Natalie:* "I notice that few of us are taking risks in talking more deeply about ourselves. I've been asking myself whether we have a trust problem."

Natalie's approach invites the others to have an open discussion about trust issues.

> *Disclosing the effect the group is having on you.* A specific type of group-related self-disclosure involves revealing how what's happening in the group is affecting you.

> *You:* "This is the third time that we've discussed sexuality, and this is the third time I've been rather quiet. My anxiety really goes up whenever sexuality becomes a group topic. I haven't tried to steer us away from discussing it, but I need some help if I'm going to talk about myself, because I feel very embarrassed."

250

In this case, the topic the group is discussing arouses your anxiety. Notice that you don't start out by trying to discuss your sexuality. Rather, you say what effect the group discussion is having on you, and you call on the other group members to give you some help in talking about yourself. Here is another example.

> *You:* "Jan, when you and Marvin talk about your relationship in the group, as you've just done, you talk very intimately. You talk about how close you feel, how you feel sexually, how you'd like to develop your friendship outside the group. Ordinarily I feel quite free to enter conversations; I feel that all conversations belong to the group. But for some reason I'm very hesitant to get into your conversations. I feel that, well, they're almost private. Maybe I'm the only one that feels this way, but I'd like to check it out with the others."

Here you feel excluded in some way, but you're not sure whether it's just you or whether others feel the same way. You reveal your own feelings of being excluded and ask whether others feel excluded too.

Revealing how you see yourself acting in this group. Another form of group-related self-disclosure involves revealing your own view of your group participation. This differs from revealing something about your interpersonal style without relating it to your participation in the group—that is, with *these* people at *this* time. Let's look at an example.

> *Rodney:* "I'm an impatient person. If I have something to say to another person, I want to say it immediately, even though it doesn't fit into what the other person is saying. Therefore, I sometimes interrupt a conversation to get my own needs taken care of. This is related to a tendency in myself to be self-centered."

Rodney is revealing something about his interpersonal style, but his self-disclosure isn't directly related to his participation in the group. Notice how the following statement differs.

> *Rodney:* "I feel ashamed of myself right now. Karen, I know that you and George are trying to mend fences right now after

251

the fight you had last meeting, but inside I'm impatient. I want the two of you to finish up your business quickly, because I want some of the group's time. I'm beginning to see that I do that all the time here. I stop thinking about other people's needs and try to get my own needs taken care of."

Here Rodney deals with the same issue, but he relates it to what's happening in the group. As a result, his self-disclosure is much more concrete. Also, his self-disclosure becomes a group-specific skill; that is, it promotes the goals of the group. Too often you may think that what's happening inside you during group meetings isn't important enough to bring up, but frequently it's precisely what's going on inside that determines how you participate in the group.

Exercise 45
Encouraging "What's Happening-to-Me-Right-Now"
Self-Disclosure

This exercise can be used from time to time to stimulate more immediate participation in the group. It can be used when the group is facing some crisis—for instance, if a member says he wants to give up because "there's no trust in the group" or if nothing seems to be happening in the group.

Directions

Stop the interaction (or interrupt the silence). Everyone takes a few minutes to ask what's going on inside themselves right now, giving special attention to what they think and feel about what's happening in the group. Each group member should ask himself or herself "What have I been thinking or experiencing in the last 20 minutes that I haven't discussed but that would contribute to the group?"

Example

Sally: "I've been bored. I haven't been attending or listening. I actually shut myself off from the group and began to daydream about the weekend I'm going to have with some friends. I'm embarrassed to tell you this."

Lowell: "Ray, I was getting angry with you. You were saying a lot about yourself, but it sounded like a monologue."

Ray: "I was getting angry, too. I thought I was saying some significant things about myself, but nobody was responding to me. I felt out in left field."

Here we can see that there were a lot of feelings that weren't being dealt with directly. These feelings were obviously affecting the quality of the group. Now that they've surfaced, the members of the group can face them.

Getting help in self-disclosure. If you'd like to disclose yourself more in the group but find it difficult to do so, use the resources of the group to help yourself do what you want to do. For instance, choose some group member you trust. Tell him or her outside the group what you'd like to disclose about yourself in the group. Once you've told this person, you'll probably find it easier to tell the others. You can also count on the encouragement and support of this person when you bring your disclosure into the full group. Remember, however, that the reason for talking over what you'd like to disclose is to get your disclosure *into the group*. It's not just a question of getting some issue handled outside the group.

Relating there-and-then self-disclosure to the group. I've emphasized focusing your self-disclosure on the here and now rather than on the there and then. You're establishing relationships with *this* group of people at *this* time. Therefore, talking in an extended way about what goes on outside the group or what has taken place in the past causes your self-disclosure to lose its impact. The question you should ask yourself about your self-disclosure is this: "What could I disclose about myself that would help me move more deeply into relationships with my fellow group participants?" Use the following exercise to help make your self-disclosure relevant to *this* group of people. The exercise is intended to help you prepare your *self-disclosure agenda* for the group.

253

Exercise 46
Preparing Self-Disclosure for the Group

Use the following format to help you prepare your self-disclosure. First, ask yourself what you'd like to disclose about yourself *at this time* in the development of your group that would help you pursue the group's goals (examining interpersonal style, establishing relationships, changing interpersonal behavior) more effectively. Make your answer as concrete as possible.

Example

You: "Since we're dealing with interpersonal relationships here, I want to talk briefly about how poorly I relate to my father. My mother is dead, and my father and I are constantly fighting. I can't please him at all. He disapproves of my major, which is history. He says it's useless. He also says that my friends are losers. And he says that my part-time job, which is counseling in a school for retarded children, won't get me anywhere either."

Notice that in this case your self-disclosure is there and then; it deals with what's happening to you outside the group.

Next, show how what you propose to disclose about yourself can be made relevant *to your group* or to your relationship with any one member. If your self-disclosure is there and then, it must, in some way, be legitimized by relating it to the here and now.

You: "Peter [your instructor], you're quite a bit older than I am. Every time I think of talking to you in the group, I tighten up inside. It doesn't make sense, but I get angry at myself and angry at you even when I just think of talking to you. Since it's unreasonable, I keep pushing my anger down. But this takes up most of my energy, and I end up practically never saying anything to you at all. I know you're not my father, but I'm still reacting to you as if you were. There's no reason for it here, but I'm afraid of you."

Here you relate your there-and-then self-disclosure about your father to your relationship to the group's instructor.

Your there-and-then self-disclosure is no longer distracting; in fact, it helps to clarify your relationship to Peter. Talking about the there and then can actually be helpful *if* you related it to the here and now. Otherwise, it turns the group into a counseling group, in which members deal with the problems they're having outside.

Taking legitimate risks in your self-disclosure in the group. As I've said before, taking reasonable risks is part of the lab experience. This includes taking risks in what you reveal about yourself. If your trust of four of your fellow group members is quite deep, while your trust of the other two is only moderately so, one way to take a risk is to reveal yourself to the whole group in the way you would reveal yourself to the four whom you trust more deeply. If you can deeply trust some of the group members, you have a basis for trust and consequently a basis for risk. Too often the self-disclosure in a group becomes too thin. The members have good skills—all the skills we've seen so far—but using good skills in a group in which the self-disclosure is too "safe" is like being a skilled artist who tries to paint a colorful picture with only white and gray paints. Many of us fear that something awful will happen to us if we reveal ourselves deeply. If there's a basis of trust in the group, however, usually the opposite is true: good things happen when group members are willing to take legitimate risks in self-disclosure. Remember, though, that all self-disclosure, no matter how risky, must still be *appropriate* in the way in which we've defined appropriate self-disclosure.

Using Responding Skills in a Group Setting

We've seen the importance of basic understanding in one-to-one conversations. It's just as important to communicate basic understanding in a group, but it's also more difficult. One reason it's more difficult is that group members seem to forget about the communication of basic understanding once they leave the one-to-one training phase of the lab. They fall back into old patterns of judging and evaluating or ignoring what others say. A high level of basic understanding in a group provides encouragement and support and deepens the level of trust. If a group has problems with trust, one of the first things the members can do is to examine how much basic understanding is provided when individual members speak. Often, everyone in a group seems to assume that someone else is going to provide basic understanding. Basic understanding is so important in the group that care must be taken to provide it *even if this must be done artificially*. In the early stages of a group, one way to accomplish this is to establish a rule: once a member speaks (for instance, disclosing something about himself or herself), basic understanding must be communicated to this person *first*, before any other kind of interaction takes place.

> *Joanne:* "There's one emotion I have a great deal of difficulty with, and that's jealousy. I get ashamed any time I think I'm jealous of anybody else. I like both you, John, and you, Kathy. The two of you have developed a really refreshing relationship here. When I see you relating so well both here and outside, well, I guess I get jealous. I don't even want to let myself experience it at all."

> *John:* "You really get down on yourself when you have to face the fact that you can be jealous."

> *Neil:* "I'm glad you brought up jealousy. I don't think I'd have the courage to bring it up. I'm right with you. I don't experience a lot of jealousy here—maybe I'm afraid even to look. But I know that it bothers me outside. I feel so petty and so small when I see myself getting jealous even in little ways."

Here John provides basic understanding. If Neil had responded first with his own self-disclosure and without taking the time to communicate understanding to Joanne, then Joanne might have felt out on a limb—wanting understanding but not getting it. *This happens too frequently in groups*, and then the members wonder why things aren't going well, why the level of trust and risk is so low. As we've seen, self-disclosure is appropriate only in a context in which the self-discloser has reasonable assurance that he or she will be understood.

It's important for you to get feedback from the other members of your group concerning how effective you are in providing the group with human nourishment in the form of basic, accurate understanding. If you're dissatisfied with yourself in this regard or find out that others are dissatisfied with you, you can increase the frequency with which you supply basic understanding in the following ways.

- *Proceed mechanically for a while.* Make a contract with yourself to respond with basic understanding a certain number of times during each group session. This procedure isn't ideal, because you want to provide understanding naturally, *but it's better than not providing understanding at all.* Once you begin to provide it more naturally, you can drop the mechanics.
- *Get a monitor.* Get one of your fellow group members to observe both the quantity and the quality of your communication of basic understanding. Better yet, let the entire group know that you feel deficient in this area. Letting others know puts you under pressure to perform.
- *Reward yourself.* There may be other forms of interpersonal behavior you like to engage in more than you do the communication of understanding—self-disclosure, for example, or confrontation or immediacy. Don't allow yourself to engage in these behaviors until you've responded a certain number of times with basic understanding. Then reward yourself with the behavior you prefer. Again, this is mechanical, but it gets results.
- *Practice outside the group.* This is a hard-core truth: *You won't respond frequently with basic understanding within the group, and you won't do so naturally, until and unless communicating understanding is part of your everyday interpersonal style.* Sometimes a group develops "specialists" in the communication of understanding, and then the others become lazy and leave this task to the specialists.
- *Take a look at how good you are at this skill.* One reason people don't provide much understanding in the group is that they aren't good at it. They've never developed this skill sufficiently to feel comfortable at it. If this is true in your case, you need further training and practice outside the group.

Using Challenging Skills in a Group Setting

The most common error people make in using challenging skills in groups is failing to use the resources of the group. Put positively, since challenging skills are "stronger medicine," one way you can be careful in their use is to get the other members of the group to help you use them well. Let's say, for instance, that Jeff confronts Ann in the group. He describes Ann's behavior and his reaction to it and invites Ann to examine her behavior. But he ends his confrontation by inviting the other members of the group to give their opinions:

> *Jeff:* "At least this is the way I see your behavior, Ann. I'm not sure whether it's just me who sees you this way, or whether the others see you this way also. I think that it would be helpful for both of us to check this out with the other members of the group."

This statement could have been worded in a number of different ways, but, whatever the wording, it constitutes a kind of perception checking. In a way, it says that confrontation is a *community affair*, that it belongs to the entire learning community.

Both the person doing the confronting and the person being confronted can call on the resources of the group to help make the confrontation productive. For instance, Ann might say:

> "I'd appreciate finding out whether others in the group see me in this way. This is a touchy area for me, and I could use some help in exploring it."

Of course, we assume that Ann has tried to make sure that she understands what Jeff is saying to her.

> *Self-confrontation and the resources of the group.* As we've already seen, self-confrontation is an important interpersonal skill. People who confront themselves, however, aren't being fair to themselves unless they call on the members of the group to help explore the area of confrontation.

> *Linda:* "I think I have the skill necessary to initiate many more conversations than I do. Frankly, I'm lazy. I get away with it here and outside by acting like a kind of pleasant introvert. That way people don't expect me to take the first step. Does this ring a bell with you people? I want to be more initiating here, and I'd like some help from you in placing this demand on myself."

Linda calls on the resources of the group in two ways: she does some perception checking ("Does this ring a bell with you people?") and also asks for help in getting herself to be more assertive.

One way to use the group's resources is to make any confrontation that occurs in the group an opportunity for self-confrontation.

For instance, if Angie confronts Norma about her lack of assertive-ness, *all* the members of the group can use this confrontation to examine the quantity and the quality of their assertiveness—not to rescue Norma but to emphasize mutuality in the group. This prevents the group from becoming a traditional counseling group in which all the members focus on the "problem" of just one. In summary, then, when any confrontation takes place, each member can ask himself or herself whether he or she should engage in some self-confrontation on the same issue. Each confrontation becomes a *theme* that all the members can profitably consider.

Immediacy in a group setting. Since a key group goal is that the members establish and develop relationships with one another, immediacy ("you-me" talk) is a most important group-specific skill. Two group members who are examining their relationship with each other can ask the others for help. For their part, the other members can offer their help when it seems appropriate to do so.

> *Naomi:* "Art, outside the group we kid each other a great deal. But here in the group we're very polite with each other. We're almost formal. I see myself as being careful with you in here. And I don't get serious with you."

> *Art:* "My experience is about the same. At one time I thought that we just weren't interested in each other. But now I'm beginning to see that we're avoiding getting close to each other. It's like both of us feared something. But this is pretty vague."

> *Marty:* "I have a couple of impressions to offer. Naomi, I see you as rather aggressive with the women in the group but much more cautious with the men. For instance, you've con-fronted Nancy several times, and you've done a good job of it. But you haven't confronted me or George or Art at all. You may be avoiding intimacy with Art, but it may be that you find more intimate interactions with all of the men harder."

Here Marty shares his impressions as they relate to the im-mediacy issue between Naomi and Art. Marty thus becomes a re-source for them, helping them to clarify their relationship. Any two members of the group, even if they're open and working hard at clarifying their relationship, can benefit from getting to see their relationship as the other members of the group see it.

Marge (to Naomi and Art): "Even when the two of you are playful outside the group, you seem to be careful not to go too far. When I really want to get to know someone, I'm often quite playful, but I'm absolutely sure that my playfulness doesn't hurt or sting. That's the kind of impression I get when the two of you play."

In this example, two members of the group spontaneously offer feedback to help Naomi and Art examine their relationship. If the feedback had not been offered, Naomi and Art could have asked for it. Calling on the resources of the group is a group-specific skill.

Exercise 47
Giving Feedback to Others on Their
Relationship

Answering the following three questions can help you prepare yourself to give others feedback on the quality of their relationship. Sample answers are given for each question.

1.   How, concretely, do I see the two of you relating?
   • "You're direct and open with each other. But you're also careful. When you bring up sensitive issues with each other—for instance, when you talk about the negative feelings you have about yourselves—you're both very tentative and supportive."
   • "Frankly, I see the two of you avoiding each other. If I'm not mistaken, there's not even much nonverbal contact. You don't look at each other much—sometimes even when you're talking to each other."
2.   Where do you two seem to be heading in your relationship?
   • "I see the two of you getting more comfortable with each other. My bet is that, if you had the time and opportunity, you'd start seeing each other outside the group."
   • "You always try to understand and support each other, but I don't think that you ever challenge each other.

It's almost like a pact: 'We like what we've got with each other. Let's not do anything to mess it up.'"

3. How does the way in which you two relate affect me?
   - "I'm stimulated by the initiative you take with each other. When you confront each other so well and respond so nondefensively, it makes it easier for me to try to do the same thing."
   - "When you get angry with each other and then just let it hang, I begin to feel the tension. I shut down and wait for the storm to blow over."

What other questions can you ask that will help you to become a resource for others who are trying to examine their relationship?

Further Group-Specific Skills

We've seen that self-disclosure, responding, and challenging skills are primary group-specific skills. We've also reviewed ways of using these skills in a group. However, this assumes that you have the assertiveness to use them. So now we'll look at other group-specific skills that will help make you an initiator, a self-starter.

Let's begin by describing once more what is meant by *group-specific* skills. Group-specific skills are ways of being a good communicator in a group as opposed to a one-to-one situation. Group-specific skills give you the ability to get group goals accomplished. For instance, a group goal in this lab is for each participant to establish some sort of personal relationship with each of the other members of the group. A skill that enables a member to accomplish this goal is a group-specific skill. This particular goal can only be achieved if the members initiate conversations with one another. Therefore, the ability to initiate conversations is a group-specific skill. Group-specific skills also include using the resources of the group to accomplish group goals, rather than having to depend only on your own resources. Therefore, when you ask other members whether they see a certain group member the same way as you do in your confrontation of this member, you're using a group-specific skill—calling on the resources of the group to accomplish a group goal.

Now let's look at some additional group-specific skills. Some of these skills are related both to one another and to the group-specific skills already described. Since communication in groups is a complicated process, some overlapping cannot be avoided.

1. Help establish a positive group climate. There are several elements that go into establishing a positive climate in a group.

- *Work.* Learning skills, examining interpersonal style, establishing and developing relationships, changing interpersonal behaviors—all of these demand work. Therefore, to present yourself in the group as a worker helps to establish a climate of work. People who aren't ready to expend effort because they don't know whether they want to live a more intense and constructive interpersonal life weigh the group down. You and the other members of your group can ask yourselves "Are we a working group? Are we putting in energy and effort?" If you can answer "yes" to these questions, then you're doing much to establish a positive group climate.
- *Learning.* Educational systems are filled with useless learnings that bore students. Work is meaningless unless it accomplishes something worthwhile for those expending the energy. You can ask yourself "Do I come off in this group as a *learner*, not in the traditional sense of somebody who memorizes information and goes through lifeless educational steps but in the sense of one whose behavior changes because of what he or she learns?" Learning means accomplishing something, being able to do today what you weren't able to do yesterday. Education can be boring; learning is exciting, even though it might be difficult. As a learner, do you take and *use* in the group what you learn?
- *Cooperation.* Cooperation contributes much more to learning than does competition. Achieving excellence in interpersonal skills is a worthwhile goal, whereas merely trying to do better than the next guy doesn't make a lot of sense. Cooperation shows interpersonal strength. If I'm insecure and have to prove myself, if I'm dependent, if I'm aggressive, if I'm self-centered, then I will see cooperation, mistakenly, as weakness. The strong, outgoing, assertive, self-confident person is actually the best cooperator. He or she sees that *everyone* gains most through cooperation.
- *Support.* By communicating basic understanding, by encouraging others in their learning (and doing so as an

equal rather than as a parent), and by recognizing the successes of others, you help to establish a climate of support. A positive group climate is a rewarding one, one that attracts people to learn. Again, self-centered, cynical, or aggressive people often see support as weakness. Support is human nourishment, and we all need it.

• *Challenge.* Support without challenge can be lifeless. Challenge that grows out of understanding and support adds life and color to a relationship or to a group.

If you're working and learning in a cooperative, supportive, and challenging way, then you're a *giver* and not just a receiver in the group.

2. Respond actively when contacted by another group member. Each of the other group members will be contacting you in order to establish and develop a relationship with you. You can add to the life of the group by responding to others *actively*. Let's take an example.

Nora: "June, you have an appealing way of confronting others. First, you confront yourself. Then you point out ways in which you think the other person is like you. When you do this with me, it keeps me from becoming defensive."

June: "Thanks for saying so. I appreciate it."

June responds, but not actively. She doesn't contribute anything to the movement of the group. Notice the difference in this response:

June: "Thanks, I appreciate your telling me. But this makes me take another look at myself. One of the reasons I confront myself first is that I'm actually scared to death to challenge others. Something in me says 'If you get yourself first, the other person won't mind as much.' By now I think that I should be able to feel free to confront you without apologizing inside myself. Often I think that I tone down what I have to say because I'm afraid of making an enemy. I wonder whether any of you see this in me."

By this response, June not only becomes a very active responder but contributes to the movement of the group. Responding in ways that add something to the group is, then, a group-specific skill.

263

3. Listen to what's happening in the group. Awareness is an important part of all the skills discussed so far. The same is true of group-specific skills. If you want to add to the movement of the group, you must be aware of what's happening in the group. This involves more than merely listening to individual members. It also means listening to interactions between members. For instance, John confronts Bill, Bill responds, and the two talk about the issue a bit. But their conversation runs out. You get the feeling that neither of them is really satisfied. Instead of letting someone else bring up some other topic, you can come into the conversation and say:

> "I've got an 'unfinished' feeling inside. You two talked a bit, but I'm not sure that either of you is satisfied with what you said."

In this case, you've been listening to an *interaction* and not just to Bill or to John. Your attention to their interaction gives both yourself and the others an opportunity to help Bill and John straighten out the issue between them.

Listening to the group also means listening to the climate or the tone of the group. At one time or another, if you're attending sufficiently, you might come to realize that there is an angry tone in the group, or that the group is peaceful after finishing some strong interactions, or that there seems to be a lot of caution and hesitancy floating around, or that a number of members are bored. This listening involves deeper understanding of what's happening in the group, because often people don't identify directly and clearly what's going on.

> *You:* "No one has said anything about being angry, yet I feel that there is a lot of anger in the group. Peter, you're saying very little. George, the tone in your voice sounds annoyed. Thelma, you're frowning. I wonder what's going on."

Obviously this kind of awareness takes work and intelligence. It's not a question of playing group psychologist or reading the "group mind." It's rather a question of working at being sensitive to the many communication cues that are constantly afloat in the group. Even the silent members are often talking to one another with their eyes, with their gestures (or lack of them), with their bodily postures. You can't zero in on each little communication cue—that would be unnecessarily distracting. But you can be aware of group themes, just

264

as you learned to be aware of individual themes in your effort to communicate deeper understanding to others.

4. Earn credit for stronger interactions. One group-specific skill is the ability to help the members of the group move more deeply into an examination of interpersonal style and more intensely into relationships with one another. But, in order to do this, you must earn credit with your fellow group members—credit that gives you the freedom to use stronger medicine, such as the skills of challenging in the group. How do you establish your credit? You establish it in three basic ways:

- *by taking reasonable risks in self-disclosure.* If you want others to talk about themselves more intensely, do it yourself first. For example, if you want to confront others, first talk about the discrepancies in your own interpersonal life.
- *by providing support through the communication of basic understanding.* After showing your willingness to deal with yourself through self-disclosure, build trust by responding frequently with basic understanding. A person who understands is in the best position to challenge.
- *by recognizing the good interactions of others.* When people do good things in the group, let them know that you're aware of it and that you appreciate it.

You: "Jim, you've taken a number of risks today. Talking about how negatively you can feel about yourself has been pretty draining for you. I appreciate your putting yourself on the line like that."

Recognizing achievement doesn't mean becoming a member of a mutual-admiration society. It does mean that you take the time to recognize others' efforts and accomplishments, however briefly. This kind of reward for good interactions has the effect of increasing such interactions in the group. It also establishes you more firmly as a trustworthy person. Of course, you must learn to do this genuinely and not just mechanically.

Doing these three things consistently and genuinely establishes your credit in the group. By doing them, you show that you respect the life of the group and that you're interested in the group's common good. You become the kind of person from whom others are willing to hear stronger things.

5. Come to the group with a relevant, concrete agenda. I've talked about keeping a log and developing an agenda or list of things to be done in the group. Making up a good agenda involves a number of elements:

- *Write down what you want to disclose about yourself.* What self-disclosure is meaningful for this group of people at this time? What can you say about yourself that will help you to explore your interpersonal style more effectively? Is there some risk taking in your self-disclosure?

  "I will talk about my fear of affectionate feelings—both in giving and receiving—and the affection I feel for Tom and Sue."

- *Write down what confrontations you'd like to engage in.* Prepared confrontations are often more concrete and thoughtful than off-the-cuff confrontations.

  "Carla was more assertive in the group for a while, but now she has slipped back into being more passive. Encourage her to act more assertively again. Describe how assertiveness makes her attractive and adds to the work and learning climate of the group."

  Don't think you must always confront. Let your confrontations grow from an understanding of your fellow group members and of what is or isn't happening in the group.

- *Write down the immediacy or "you-me" issues you'd like to face.* Try to come up with issues that make a difference in your relationships with the other members of the group.

  "Loretta and I are both very active in the group. I think that I'm competing with her. I also think I'm trying to outdo her because I'm male and she's female."

If your group has seven members, and if all seven of you bring good, concrete agendas to the group each meeting, you won't be sitting around wondering who is going to say something next. Rather, you'll have to fight to get time for your agenda, and the group will be a busy, working group.

6. Initiate conversations with others. Initiating conversations is one of the most important group-specific skills. We've seen that it's a group-specific skill to respond *actively* when others start conversations with you. But that isn't enough. Being assertive means starting conversations with others, making contact with them before they make contact with you. Assertiveness makes the group a lively place to be. You're already familiar with ways in which you can initiate conversations with others:

- *Self-disclosure:* "Zelda, I'd like to share something about myself with you and get your reaction."
- *Communication of basic, accurate understanding:* When somebody says something about himself or herself without addressing his or her self-disclosure to anyone in particular, you become an initiator if you volunteer to respond with basic understanding.
- *Confrontation:* "Something you do here, Tim, has begun to bother me a bit lately. I'd like to talk about it."
- *Immediacy:* "Theresa, we haven't talked for a couple of weeks now, and I feel some rumblings that things between you and me aren't entirely right."
- *Encouragement:* "Vince, I'm really glad you've worked out your anger with Jennifer. I get the idea that we're all breathing easier."

Groups bog down and become stale and boring when the members adopt the attitude "Let George do it." If each member has the attitude "I'll do it," the group will be full of energy. Ask yourself this question: "Have I initiated conversations with every member of the group over the past couple of weeks?" If you find that you do start conversations with some but not with others, then it might be useful to examine why. This examination should lead to some immediacy issues.

I've emphasized that an assertive person is one who gets his or her needs and wants met without stepping on the rights of others. A passive person doesn't get his or her needs and wants met and even allows others to violate his or her rights. An aggressive person gets his or her needs and wants met but violates the rights of others in doing so. In the group the contract spells out some of your basic rights. For example, you have the right to examine your interpersonal style. If you don't take the opportunity to do so, you're a passive member of the group. If you hog the time of the group, you're an aggressive member. If you examine your interpersonal style and help others to do the same, you're an assertive member.

You also have a right to challenge others *if* you first earn the credits you need to do so. If you communicate understanding to others and then refuse to challenge them, you're a *passive* member. If you challenge others in a rough, uncaring way, or if you challenge others without first earning the right to do so through your own self-disclosure and communication of understanding, you're an *aggressive* member.

What are some of the other rights you have in the group? This whole human-relations-training program is designed to make you more active or assertive in your interpersonal life. You may say to yourself "I don't want to be more active in my interpersonal life." What is asked of you in the laboratory is to *experiment* with being more active. See some of the possible rewards for being more active. When the lab ends, you can decide how active you want to be in your everyday interpersonal life.

7. Ask for feedback. One excellent way to initiate interactions in the group is to ask others for feedback on what you say in the group, on your interpersonal style, or on the quality of your confrontation. Recall that an important group-specific skill is *using the resources of the group.* If you aren't getting enough feedback, ask for it; it's your right. If you wait for others to give it to you, you're being passive. If, after waiting, you explode and tell the others off, saying that they don't care about you, you're being aggressive. If you make reasonable requests for specific and concrete feedback, you're being assertive.

> *You:* "I've been trying to increase the number of times that I volunteer to respond with basic understanding. It's not just a question of how many times, though. I'm interested in finding out something about the quality of my responses. Could I get some feedback on what I'm doing right and what I'm going to have to improve?"

If you learn to make legitimate requests, you add energy to the group.

8. Enter ongoing conversations. In a training group, there are no private conversations. Many group members deny this in practice by refusing to interrupt a conversation two other people are having. Assertive group members ask themselves, while listening to the conversations of others, "Do I have anything to contribute to this conversation?" Assertive members also observe the following practices.

268

• *Avoiding disruption.* There is a huge difference between entering a conversation and taking it over. If you barge into a conversation and refocus it on yourself, you're disrupting the conversation and being aggressive.

> *Murray:* "What you two are talking about reminds me of what happens when my wife and I get into a fight. Last week we were having what I thought was a quiet conversation when all of a sudden . . . ."

Here Murray doesn't contribute anything to the ongoing conversation. He doesn't help those talking to clarify anything. He just barges in and takes over. He steps on the rights of others in order to get some of his own needs taken care of.

• *Using interjections.* If two group members are having a conversation and you make some very brief statement indicating that you're attending and listening, you're using interjection.

> *Lonnie:* "Fred, you said you've come a long way in how you feel about yourself. My guess is that in many ways you feel a lot better about yourself, but I'd like to hear more specifics."

> *You:* "I'd really like to hear that too, Fred."

Your brief statement is an interjection, a sign of your presence and interest. It goes without saying that interjection can be overused, making you appear to be a busybody. Used sparingly, however, it can add to the energy of the group.

• *Showing involvement.* Involvement means that you feel free to join any ongoing conversation if you feel that you have something to contribute that will not disrupt what the people having the conversation are trying to accomplish. It's your right to involve yourself, since there are no private conversations in the group. You can contribute to an ongoing conversation by:

• *communicating basic understanding* to either or both participants. This may help them clarify what they are trying to say to each other.

• *disclosing* something about yourself if it relates to what the other two are saying to each other and helps them achieve their goals.

269

- *confronting,* if confrontation is appropriate and if you do it caringly: "Tom, your slouching posture makes it hard for me to see that you're interested in what Bill is saying to you. And Bill, you're not even looking at Tom. I'm distracted by what you two are doing to each other nonverbally."
- *using immediacy:* "Dianne, I don't want to gang up on you, but I'd like to confirm what Jenny is saying to you. When you become passive, I don't want to cuddle you, either. I want to shake you. I want to relate to you as an adult, not as a little girl."

The cultural norm seems to be "Don't interrupt an ongoing conversation. Be polite." We'll change that in the group to "Don't *disrupt* an ongoing conversation, but do involve yourself in it if you have something to contribute. Be assertive."

9. Promote dialogue instead of monologue in the group. Groups in which people give long speeches, sermons, or monologues are usually dull and lifeless. Monologues sap the energy of the group. The group-specific skill of encouraging dialogue has two parts. First, speak briefly and in a way that lets others know that you expect a response. Notice the difference between the following two statements.

> *You:* "I find affection difficult to express. One way I show affection is to kid around with others. I can say almost anything to others if I say it in an outlandish way. For instance, I can exaggerate and say 'Hey, I haven't seen you for almost a week! Why do you do this to me—I mean, making my life bleak and unhappy for a whole week!' I get my message across: it's affection, but nobody can accuse me of being mushy."

Here you reveal something significant about yourself and your interpersonal style, but nothing in what you say indicates that you expect a response. Nothing in your statement indicates that what you're saying refers to *this* group of people. Now consider the following statement.

> *You:* "I've been thinking of how differently we express affection here. For instance, I do it by saying things in an outlandish way. I've said to you, Jan, 'Call me up this week, or I'll go

270

mad.' You get the idea, obviously. But I have to be pretty indirect. I think I come across as fairly unemotional and therefore as not very affectionate, but I'd like to check that out with the rest of you. Peter, you can say 'I like you' to someone quite warmly and naturally. I don't think I do that well at all."

In this statement, you bring up the *theme* of the expression of affection. You ask for feedback, and you give some feedback of your own on this theme to Peter. Your examples are related to the group. Your whole style of presenting yourself is designed to get some discussion going in the group and to get some feedback for yourself. And yet you didn't give a long speech about yourself.

Second, to encourage dialogue, don't let others give long speeches. If they don't relate what they're saying about themselves to you or to the group, then you can do it.

You: "Greg, you're talking about how difficult it is for you to trust others. I sit here wondering just how much you trust me—or any of the others, as a matter of fact. I guess I have some feelings about how well we're doing in building trust in our group."

Here you interrupt a monologue and give a group focus to what the other person is saying. Just as one way of making your self-disclosures more concrete is to relate them to the people in the group, so too you can help others make what they say more concrete by asking them to relate what they're saying to you and to your fellow group members.

10. Use the present tense instead of analyzing the past. There is one sure sign that the members of the group have stopped interacting with one another and started analyzing what they've already done, and that sign is the use of the past instead of the present tense. There's a natural tendency in all of us to sit around and talk about what *has* happened. Perhaps this tendency is even stronger when we sit around in groups. Consider the following brief dialogue.

Henry: "Ron, last week, after I confronted Joy, I think that you got angry with me, but you didn't say anything. I kept looking at you out of the corner of my eye, and you wouldn't even look at me. My guess was that you were so angry that you couldn't let yourself say anything."

Ron: "You're absolutely right. I went home last week angry and confused. I thought you and Joy were close, but I didn't see how your confronting her about her inability to be a fun person in the group had any meaning. It seemed to me that you were just attacking a close friend, and about a silly issue at that."

Joy: "I felt so ashamed about crying. I felt humiliated. I actually felt like a little girl being scolded. I knew you were all steamed up, Ron."

First of all, most of what is being said here should have been said *at the time it was all happening.* This group postmortem is very safe and accomplishes very little. The participants are talking about history; they're not really working out interpersonal issues. They're analyzing, not interacting. It's a group-specific skill to turn this analysis into interaction and relationship building. Notice the difference in the following interaction.

Henry: "Ron, I've got a pretty good idea that you were furious with me last week when I was confronting Joy. Can we talk about it?"

Ron: "I'm still angry. First of all, you tend to confront what I would call 'safe' people. Joy isn't the type to get back at you. You don't confront me at all, and I think that I resent that."

Joy: "I've thought about what happened last week. I don't think that you pick on me, Henry, but I do think that I'm easy for you to confront. I don't speak up. I'm not at all assertive when I'm being confronted. I don't like that in myself, and I want to do something about it."

In this conversation, Henry, Ron, and Joy use the past as a point of departure. They don't spend useless time analyzing last week's interaction. Rather, they take up where they left off. They get right into the middle of things.

If you're *listening* to what's happening in the group, you'll quickly become aware when people stop interacting and start analyzing, when the past tense takes the place of the present. That's the moment to challenge others to return to the present.

> *You:* "Joy, I think I know what happened between you and Henry last week and how you felt. I'm wondering how you want to relate to each other right now."

11. Challenge silent decisions that make the group less than it could be. Sometimes, without even knowing it, the members of a group enter into a private conspiracy and make decisions that rob the group of energy and vitality. Usually these silent decisions make the group "safer." Let's consider an example. Bert begins talking about some of the sexual feelings he has toward a couple of the other members of the group. These members hardly respond at all. They certainly don't respond actively. No one else in the group takes up the theme of sexual feelings. As a result, this theme dies. A couple of meetings later, Lisa mentions something about sexuality, and she too receives little if any response. Note what's happening in this group: the silent majority is saying, by their very silence, "We don't talk about sex here." A *silent decision* has been made.

Groups tend to make these silent decisions, and once they are made, they're very difficult to reverse. Since such decisions often make the group too safe and rob it of energy and vitality, they need to be challenged before they become "law." For example, your group might make a silent decision not to interrupt when two people are talking. One or two people try to join ongoing conversations but are ignored. It isn't long before the silently made rule weighs heavily on the group: "We don't interrupt conversations in this group. We wait until people are finished." Challenge such a decision before it becomes law.

> *You:* "I've tried to get into conversations a couple of times tonight, but I feel that I was ignored. I wasn't going to push my way in, but I'd like to get the feelings of other members on this. I don't want to become a silent observer of private conversations. I feel left out and bored when that happens. I don't want to disrupt conversations, but I want to be allowed to join them. I'm angry right now, and I feel left out."

Here you're being assertive about your rights as a group member. You put the issue to the group directly and invite them to discuss it. If the group wants to make a decision to limit the group in some way, let the decision be public, and let it be made after some discussion.

12. Encourage mutual rather than helping relationships. I've already cautioned you a number of times against playing psychologist or becoming either a helper or a client in the group. You and your fellow group members will avoid playing these roles if you are careful of maintaining mutual relationships. There are various ways of encouraging mutuality and discouraging helping behavior.

• When you disclose some difficulty or problem you have in interpersonal relating, ask others if they share the same problem in any way. Let your "problem" become a group theme for a while. Refuse to let the others make you a client while they become helpers. You can get a lot of help without turning the group into a kind of group therapy in which each person takes a turn playing the patient.

• When you confront somebody on an issue, add some self-confrontation—not to soften your confrontation but to make it more mutual.

> *You:* "When you're about to confront somebody, Lenny, you often start with a long apology. You just did it now with Martha. I apologize too much for what I do here. I guess that makes me sensitive to self-apology in you."

In no way do I suggest that you should make something up about yourself so that you can appear to share something with the person you're confronting. Rather, when you do confront, remember that your confrontation can become a self-confrontation *theme* for all the members of the group.

• When you communicate basic understanding to another person, go on to share something about yourself.

> *Terry:* "I've been in this group almost ten weeks now. It struck me this past week how shallow my relationships here still are. I say to myself 'What have you been doing?' It's what I'm *not* doing. I talk about myself a lot, and I listen fairly well and respond, but I practically never do anything to develop any relationship in here further. The work I do in here doesn't seem to get me closer to anyone. I'm at a loss."

> *Brad:* "You sound surprised and maybe even a bit embarrassed. Ten weeks and still no solid relationships! As you

274

were talking, I began asking myself the same question. What deeper relationships have *I* developed here? This sounds like a theme we could all take a look at."

Brad does more than merely respond with basic understanding. He joins in on the theme—lack of relationship establishing in the group—that Terry has presented. If you constantly communicate basic understanding without sharing yourself, you'll begin to sound like a counselor rather than like a fellow learner.

• When you're confronted, examine the issue with which you're confronted, but then invite others to do the same. In this way you avoid becoming a patient. If you assume either the role of counselor or the role of client in the group, you may well be resented.

13. Learn to say it in the group: Confide to the group what you would confide to someone close to you outside the group. Sometimes a married person will leave a group experience, go home, and tell his or her spouse about how he or she *really* feels about what's happening in the group. Such a person isn't necessarily violating confidentiality, since he or she doesn't use names and the spouse doesn't know anyone in the group or ever come in contact with any of the group members. The point is that the trust level of a group can be measured by the extent to which you're saying *in* the group what you used to say *outside* the group. One day I overheard two group members talking about a third group member. They were both saying how anxious they became when they tried to talk to her, because she was a very controlling and demanding person. Unfortunately, neither of them had brought these feelings up in the group, and therefore their behavior toward her was less than genuine.

14. Discuss the difficulties you have in involving yourself with the group. A final group-specific skill is to be open and honest about whatever difficulties you're having in being an involved group member. Many of us have fears of appearing incompetent in front of other people. Therefore, we're afraid to explore our difficulties with others. If you realize what *strengths* you have in the group, it might be easier to consider your difficulties against the background of your strengths.

You: "I think I work hard in the group. I always attend to others, and I work at responding well. But I'm still dissatisfied with my level of self-disclosure. Sometimes I tell myself

275

that I trust the people here and that I'm willing to discuss almost anything that relates to the group, but maybe I still don't know how to get myself across to others."

All of us from time to time have difficulty involving ourselves with the other members of a group. Such difficulty isn't an indication of deep dissatisfaction with the group or of a lack of motivation. In fact, admitting the difficulties we're having in getting involved is itself a sign of being motivated.

A Checklist of Group-Specific Skills

This checklist is a set of goals for group involvement. If you're having trouble becoming an effective group communicator, use the checklist to help pinpoint the behaviors that are most difficult for you.

A. *Self-presentation, responding, and challenging skills as ways of establishing and developing relationships in the group*
   1. *Self-disclosure skills*
      - Disclose what's going on inside you as you participate in the group.
      - Disclose how you feel about your own participation.
      - If you have a hard time disclosing yourself, ask for help, or get someone to monitor the quality of your self-disclosure.
      - Relate there-and-then self-disclosure to the here and now of *this* group of people.
      - Prepare your self-disclosure.
      - Learn to take appropriate risks in self-disclosure.
   2. *Responding skills*
      - Attend and listen carefully to all the interactions in the group. Monitor your posture and the messages you're sending by means of your posture.
      - Keep your communication of basic understanding at a high level, even if you must do it artificially for a while.

3. *Challenging skills*
   - Use self-confrontation as a way of leading into confrontation.
   - When you confront, check your perceptions out with the other members.
   - Periodically check with each of your fellow group members how your relationship building is going (relationship immediacy).
   - If you're having trouble communicating to another member, ask the other members for help.
   - Give others feedback on the quality of the relationships they're establishing in the group.

B. *Further group-specific skills*
   1. Help establish a positive group climate by working hard, trying to help others learn, and being cooperative and supportive.
   2. Respond actively when contacted by others. Cooperate in exploring the issues they bring up.
   3. Learn to listen to what's happening in the group and to the way in which individual interactions affect the group.
   4. Earn the right to challenge others by being a spontaneous self-discloser and by providing basic understanding for others.
   5. Bring a concrete agenda to each meeting.
   6. Be assertive in the group—start conversations, disclose, give feedback.
   7. Ask for feedback and whatever else you need to be an effective group member.
   8. Join and contribute to ongoing conversations.
   9. Engage in dialogue rather than monologue, and demand dialogue from others.
   10. Avoid analyzing the past; deal with the past only insofar as it has impact on present behavior.
   11. Challenge silent group decisions that rob the group of risk-taking and vitality.
   12. Don't let yourself or others fall into the roles of helper or client; encourage mutuality.
   13. Learn to confide thoughts and feelings that involve risk to the members of the group rather than to trusted friends outside the group.

277

14. Discuss the difficulties you have involving yourself with the group. See what stands in the way of your taking responsible risks.

Social Influence

Most of us like to see ourselves as relatively free and independent people. However, one goal of human-relations training is to come to a deeper understanding and appreciation of *interdependence*. As you know, being *dependent* means counting too much on relationships with others. If you're dependent, you find that you can't do something unless some other person does it. Or your opinion of yourself depends on the opinions others have of you. You think that others are better than you are. You count heavily on others to prop you up. Since being dependent is not an attractive trait, people sometimes try to hide their dependence by pretending to be *very independent*. They try to knock other people over the head with their so-called independence. They come across as aggressive. This aggressive hiding of dependence is technically called *counterdependence*. Interdependence means that you're neither openly dependent nor secretly dependent (counterdependent). It means you're not merely independent. It means that you're in possession of yourself and capable of being independent but that you *choose* to surrender some of your freedom in order to relate more deeply to another person. The give-and-take of more intimate relationships calls for interdependence rather than mere independence.

If you and I are interdependent, then we can admit that we influence each other in various ways. We can call the various ways in which people affect one another *social influence*. Social influence is part of everyday living. For example, when I show care for others, they are often influenced to like me, respect me, and cooperate with me. On the other hand, when I'm cynical with people, they're often influenced to avoid me and perhaps to fear me. Even when I don't act (especially in situations in which I might be expected to), I influence people. For instance, my silence at a meeting influences other members to think of me as incapable or unconcerned and perhaps to deal with me as a "problem." Failing to communicate basic understanding to others might make them wonder whether I care about them or am really "for" them. In summary, we're always having impact on one another whether we want to or not and whether we realize it or not. In the group we face the fact that we do influence one another, but we see this not as manipulation but as an open, growthful process. I want to make an impact on you (for instance, through confrontation, immediacy, and understanding), and I want you to have an impact on me. I assume you think the same way, since you agree to the same contract. However, the ways in which we influence one another are, as much as possible, aboveboard. In that way, we influence one another without manipulating or controlling one another.

Learning both one-to-one and group-specific skills makes you open to legitimate influence *and* helps you ward off attempts on the part of others to manipulate or control you. You cannot learn these skills well without becoming more assertive, and assertiveness helps prevent others from controlling you. Skills training can give you a greater sense of competence and increase your appreciation of yourself. You become less dependent and are freed from the need for *excessive* approval from others. On the other hand, listening more intently to others and in a nondefensive way opens you up to being affected by what they have to say. Social influence, then, is in itself neither good nor bad. It depends on how it is exercised and on what the reasons for it are. But since social influence is part and parcel of everyday life, we're better off if we understand the process and know how to use it without violating our interpersonal values.

In the training group, you're a member of a *learning community* in which cooperation, rather than competition, is called for. One of the assumptions of a learning community is that its members care for one another. Caring, then, dictates the ways in which you and your fellow members influence one another. Ideally, your group will become a "modeling network," which means that you can take as models people who do well—for instance, people who are good at both one-to-one and group-specific skills. One person may model one skill or quality that you like, while other people in the group model another. Without losing your own identity, without becoming dependent, and without becoming a slavish imitator, you can begin to add what you like in others to your own interpersonal style. Obviously, you can also become this kind of model for others. The group as a modeling network is an example of a good use of social influence.

The Open Group

In the last chapter, a number of positive ways of giving yourself to the work of the group were suggested. This chapter discusses some of the ways in which group members can run away from the work of the group—and what to do about them. The chapter also looks at five levels or degrees of participation in a group.

Defensiveness: Dealing Constructively with Flight Behavior

Putting all the skills we've been talking about together in an open group experience isn't easy. It takes a great deal of concentration and work. People, being human, sometimes flee the work of the group. A tendency to be lazy isn't necessarily a sign that someone doesn't care. Even a person who is highly motivated to give himself or herself to the work involved in the lab experience will tend to resist the work when it becomes anxiety arousing and demanding. What happens in the group—getting to know oneself better and getting closer to others—is sometimes threatening. A person who feels threatened will likely become defensive.

In this chapter, we'll look at some of the main ways in which both individual members and the entire group can avoid some of the work and the responsibilities of the group. No attempt is made to discuss *all* the ways there are of running away—people can be very creative in running away from things they don't want to face. Since flight from work can be very complicated, some of the classifications of flight described in this chapter overlap.

The Individual in Flight

Here are some of the ways in which *individual members* of the group can be in flight from the group's work.

281

Boredom. Boredom, it has been said, is an insult to yourself. In the group, a bored member usually has let himself or herself become passive. A person who is bored sees himself or herself as a *victim* of what's happening in the group and tends to put the blame "out there," saying that the interaction isn't "interesting." A bored person, then, is one who has given up taking the initiative in the group and is just letting things happen. When confronted, he or she dishes up excuses that are a bit lifeless, realizing to some degree that he or she deserves to be bored:

- "I just couldn't get into the discussion."
- "I didn't say anything because nothing was happening."
- "Nobody else was really involved in the group."
- "I couldn't seem to get started."
- "Bobbie and John were going on, but they really didn't seem to be saying anything to each other."

A bored person is a burden for the group. Since such a person is really not attending, he or she becomes a distraction. People will notice someone who is bored, even if they don't say anything. Eventually they'll feel that they have to "deal with" the bored member.

Randy: "Val, you seem really out of it this evening. What's going on with you?"

Val: "It's not me. Nothing's happening here to get very excited about. I've been bored for half an hour."

Val's response is an aggressive one. She has moved from passivity (letting herself be bored) to aggression (lashing out because she feels attacked).

What should you do if you feel bored in the group? Your first task is to be aware that you're slipping away from the group, that you're *allowing yourself* to become bored. *In other words, you take responsibility for your own emotions.* The best cure for boredom is becoming an active member in the group. Like some other negative emotional states, boredom only gets worse if it's not faced *immediately.* For instance, if Jeremy and Barbara are being vague and general and long-winded in discussing their relationship, and if no one in the group is doing anything about it, you can avoid boredom by entering into their conversation in a constructive way.

You: "I've noticed that my attention is beginning to wander. It would help me, Jeremy and Barb, if the two of you could become more concrete. For instance, both of you say that you're uncomfortable with each other. I'm not sure what each of you means by 'uncomfortable.'"

Here, instead of being passive or aggressive, you become assertive. You have the right to join Jeremy and Barbara's conversation, and you have the right to ask them to become more concrete. You can also offer yourself as a resource to help them become more concrete with each other.

Thus, the best way to handle boredom is to involve yourself with others in such a way that you avoid becoming bored. However, if you find that you're *often* bored when others are talking, you may ask yourself how motivated you are to get more deeply involved with others. If being rather uninterested in other people is part of your interpersonal style, perhaps you might want to discuss this issue. It is, of course, related to something that was discussed in Chapter 1—the relationship of values to human-relations training. Perhaps deeper involvement with others isn't a value for you. Still, in the lab you're asked to experiment genuinely with deeper involvement.

Finally, a bored person often feels compelled to defend himself or herself when challenged for being bored. However, the group should avoid long debates over who is responsible for anyone's boredom. The bored person is responsible for his or her own boredom. Debate in this case is a smoke screen.

Talking outside the group. If you're bothered by an issue that relates to the group, but you deal with it outside, you drain off energy from the group itself. For instance, let's say that you and Darlene don't get along very well in the group. One evening, after a group meeting, the two of you have coffee together. You decide to spend time over the weekend talking about your relationship. You straighten things out in your talks. You return to the group the next week and say nothing about your talks, but now you relate to each other warmly and openly. Everyone in the group is confused or distracted, because they can't account for this new behavior.

I'm not implying that you shouldn't talk to other members of the group outside of group meetings, but, if you do, you should let your fellow group members know in some summary way what has happened. Let's take another example. Suppose that you and one of the other group members are very close friends outside the group.

You don't mention this in the group, but you send a lot of distracting nonverbal messages to each other. Of course, you and your friend don't need to discuss in the group everything that goes on between the two of you, but you can't pretend that you're just getting to know each other. The other group members should have some idea of what goes on outside the group *to the degree that it affects your relationship inside the group.* In general, it's sound practice to let others know what has happened in your life between meetings that affects the way you participate in the group.

Psychologizing. I will only summarize this form of flight here, because I've been mentioning it all along. You're in flight through psychologizing if you:

- make yourself into a helper or counselor,
- make yourself into a client or patient,
- use a lot of psychological jargon ("You have an inferiority complex, and you're using a lot of defense mechanisms"), or
- spend a lot of time looking for insights into your personality instead of examining the behaviors that form your interpersonal style ("I think part of my shyness is related to the fact that my parents gave much more attention to my older brother").

The remedy for psychologizing is pursuing mutuality in all the ways I've already described.

Playing the director. It's possible for someone to appear to be very active in the group without actually getting deeply involved with others. I call this kind of person the Director. The Director does his or her best to see that others get involved but in the process forgets to get involved himself or herself. The Director moves safely into ongoing conversations but doesn't initiate them. The Director asks a lot of questions ("Paul, are you really satisfied with your interpersonal style?"), comments on what's happening in the group ("I think we're tied up right now because of the hostility Jim and George just expressed to each other"), and checks out the feelings of other members ("Jane, how do you feel right now?") but never gets very involved either in self-disclosure or in confrontation or immediacy.

People who don't "need" skills training. In an ideal family/educational system, many of the skills discussed in this book would be modeled by parents and taught in grammar school. The fact is that they are not. Sometimes, however, people come to this kind of laboratory experience vaguely feeling that they should already have all of these skills. Consequently, they feel embarrassed about having to learn skills that "should have" been learned at an earlier age. Whatever the reasons, some lab participants who actually *don't* have the skills I've outlined flee the work of the laboratory by implying in one way or another that they don't *need* skills training. They may say that they already possess the skills or that they are already quite effective in interpersonal relating.

My response to such claims is that, first of all, all of us could use a skills check-up from time to time. People who think they already have these skills can use the lab experience as a kind of extended check-up. If they find that they've slipped in this or that skill, they can use the experience to polish up their skills. Most of us, however, find that we either don't possess these skills or don't use them naturally and frequently in our interpersonal relationships. We feel awkward in learning and practicing them. Part of the spirit of a learning community is to be aware of feelings of awkwardness and to understand them in one another. It's a great mistake to express a punishing, "show-me" attitude toward people who feel that they possess certain skills when in fact they don't—to say, for example, "If you have all of these skills, then why aren't you using them here?" It's much better to try to understand why a person might be resistant to the skills-training process. However, a person who continues to resist learning skills is violating the contract of the group and is in flight.

Cynicism. A cynic is a person who sets himself or herself apart from others and in effect sneers at the possibility of sincerity. Cynics take a condescending attitude toward many of the dimensions of interpersonal living. They don't believe that people can really care for one another. They believe that, on the contrary, most of us are motivated only by self-interest. Therefore, they tend to scoff at supposedly sincere, noble, or tender human interchanges. Now we could hardly expect to run into pure cynics in human-relations-training laboratories. However, a bit of the cynic lurks in many of us, and one way in which we can interfere with the training process is to let the cynic within us get the better of us at times. If I'm uncomfortable with closeness and intimacy, I might use cynicism as a defense against being intimate. I might make fun of tenderness and closeness and sincereity instead of just admitting that I feel awkward and embarrassed when I try to get closer to others. The remedy for my defensive cynicism is to deal with my uncomfortable feelings, not to wait until I'm confronted in the group for being cynical.

Rationalizing. To *rationalize* means to substitute more acceptable explanations for my negative behavior instead of to face the real reasons for it. For example, if I'm failing to attend to others in the group, I may say that I've had a tough day instead of just facing the fact that I've let my mind wander. The former explanation is more acceptable; it makes me appear more noble. We use rationalizations to put the blame for our failures anywhere else but on ourselves. This is understandable, but it can also stand in the way of our progress in the lab. What are some of the rationalizations that are commonly heard in human-relations-training laboratories?

- "This lab isn't real life. I can't be expected to be my real self here."
- "I'm just too exhausted after a day's work to really get into things here."
- "I really don't know what's holding me back."
- "I really don't know what I can do to improve my interpersonal relationships."
- "No one else is keeping the contract well."
- "I'd be doing a lot better if I were in that other group."
- "The main difficulty is that the instructor and I don't get along."
- "All this structure is artificial. I could really move if things were more natural."
- "Labs like this shouldn't take place in a classroom. We should be in a more comfortable room with pillows."
- "There are too many skills, too many things to be learned all at once."
- "I'm really not the interpersonal type."
- "I don't talk about myself in the group because nobody is really interested in me anyway."
- "I'm quiet because others say what I want to say, and they say it first. So I end up with nothing to say."
- "I don't confront you because I don't want to hurt you."
- "We don't really have enough time to establish relationships."
- "I don't try being more assertive because I end up being aggressive."
- "I know I don't disclose a lot, but you can ask me whatever you'd like to know."
- "I know I don't communicate a lot of understanding to others, but that's just not a part of my style."

286

I'm sure that by now you could make up your own list of rationalizations. What would your list be like? The solution to rationalizing is confronting yourself, inviting yourself to examine how you're handling the difficulties of the laboratory. If you can admit the difficulties you're having and use the resources of your learning community to help yourself move beyond these difficulties *early in the life of the group,* then your need to rationalize will be less. Such "confession" in the lab might or might not be good for the soul, but it's certainly good for the training process.

Silence. Some people have tried to defend the silent members of training groups, saying that such people could be learning even though they aren't participating actively. A rather silent member of one group had an interesting rationalization. He said that he was *learning* inside the group and *doing* outside the group. Nevertheless, although a silent person might well be learning something in the group, he or she isn't *contributing* anything. The contract calls for us to be contributors and suggests that we really learn through contributing.

I read once about a girl who participated in a 40-hour marathon group experience. For the first 15 hours, she sat next to the wall, saying nothing. When the leader of the group asked her what was troubling her, she said that she had paid her money and was waiting for him to do something with her. This incident, of course, says much about group experiences that lack defined goals and structures. But even in groups governed by a contract calling for active participation, some members take refuge in silence. While it's true that quality of participation is, absolutely speaking, more important than quantity, there is still a point at which lack of quantity is damaging to the overall quality of a person's participation.

A silent person, perhaps without wanting to, can also become manipulative. His or her silence begins to bother the other members of the group. They ask what's wrong. They try to coax the silent person into participating more actively. In summary, a silent member begins to take up the energy of the group—energy that could be put to better use.

The silence of any member should first of all be understood, if possible. Does his or her silence indicate fear? Boredom? Lack of interest? A long-standing problem in interpersonal style? However, it's not productive to use too much of the group's energies trying to deal with a silent member. Like any other member, he or she is expected to deal with himself or herself. Otherwise the silent person takes on the role of client or patient.

287

Humor. Humor is a two-edged sword. It can be used to lighten the effect of confrontation, but it can also be used to run away from the work of the group. A genuinely humorous person can often get a confrontation across in a lighthearted way but still make the confrontation serious and meaningful. On the other hand, some people, when things get too tense, dissipate the tension with humor, failing to realize that a reasonable amount of tension can help keep people working toward the goals of the group. Whenever either individuals or groups adopt humor as a *consistent* part of their style, it is no longer serving a useful function, and it needs to be confronted.

> *Carolyn:* "Ned, I see you as a genuinely funny person. Humor is one of your strengths; you're good at it. But sometimes I see you using humor here when you're in an interaction that's tense. That is, you can use humor to keep people at arm's length. I think I see that happening between you and me. You can disarm me so easily with your humor. Frankly, I think I'm too humor*less.* Maybe we could talk about it."

Lack of directness. There are a variety of ways in which lack of directness can become a form of flight in the group.

• *Speaking in generalities.* One way of speaking in generalities is to use substitutes for the personal pronoun "I." Some common substitutes for "I" are "you," "one," "we," and "people." In the following examples, John is talking to Harold about the difficulty he's having in clearing up a misunderstanding with Harold. Notice the difference in the directness of the following two statements.

> *Statement A:* "People get scared when they think they should talk about what's bothering them."

> *Statement B:* "Harold, I'm really afraid to talk about the misunderstanding you and I had last meeting. My guts are churning, because I think you might still be angry."

If John uses Statement A, he's beating around the bush. He's so indirect that Harold and the other members of the group probably won't know what he's talking about. If he uses Statement B, he's directly facing the issue that's bothering him. Statement A is nonassertive; Statement B is assertive. High-level communicators value directness and assertiveness in interpersonal relationships. It's possible to be direct without being aggressive or tactless.

288

• *Asking questions instead of making statements.* Another form of indirectness is to ask a question when you really want to make a statement. For instance, you see someone having a hard time in the group, and you ask:

"How do you feel right now?"

This is almost always a poor question. It's a cliché question in human-relations training. This question, like others, can often be turned into a direct statement. For instance:

"The way your head is hanging makes you look depressed to me. You seem sad, but I'm not sure why. I'd like to know what's going on."

Here you're assertively sharing your impressions and indicating your interest instead of hiding all of this in a question.

• *Waiting for the "right" moment.* You're in flight if you're putting off things you'd like to discuss because you feel that it isn't the "right" time. One sign of this form of flight is that you keep putting items in your log and agenda but never get to them in the group meetings. People who keep waiting for the "right" moment say things such as "I didn't want to interrupt," or "Peter said what I wanted to say," or "The group had moved on to other issues," or "I thought that you wanted a rest, since you'd been the focus of the group's attention for a while." One way of finding out in an assertive way whether it's the right time or not is just to ask. For instance;

"Janet, you've heard quite a bit today about your lack of assertiveness in the group. I'd like to share my feelings about you, but I'm not sure whether this is the right time."

If Janet says that she'd rather wait, you can respect her wishes. However, she now knows that you'd like to share with her the feelings you have about her.

• *Seldom using a person's name.* Some group members almost never use the first names of their fellow group members when talking to them. Using a first name is a form of intimacy or contact. It's a way of touching others. If you discover that you seldom, if ever, use another person's first name in talking to him or her, it's time to ask yourself how closely you're relating to this person. If, when you do use another person's first name, you feel funny or phony or uncomfortable, that's also a cue to examine the quality of your relationship with the other person.

• *Talking about a person rather than to the person.* Consider the following example.

> James cries.
>
> Mary Jo grows silent and looks at the floor.
>
> Bill watches Mary Jo and points out her silence and her non-verbal behavior of watching the floor.
>
> Mary Jo responds to Bill by talking about her feelings about James: "When James started to cry, I wanted to say something but just couldn't."

In this example, Mary Jo avoids direct communication with James in two ways: first by not talking to him when he was crying and then by talking to Bill about James instead of directly to James. Here it wouldn't be aggressive to say to Mary Jo "Perhaps you could talk directly to James instead of to Bill."

• *Avoiding sharing hunches and perceptions.* In this common form of flight, you avoid taking the risk involved in sharing your hunches about others and your perceptions of their behavior. For instance, you feel that Bob is much more anxious than he's willing to admit and that he confronts others before trying to understand them, not because he's thoughtless but because he's nervous. This is your hunch, but you never share it with him and therefore never give him the chance to do anything about it. Instead you just wince inside when he confronts others carelessly.

Low tolerance for conflict and emotion. Often in a group there are a couple of participants who have a low tolerance for conflict or strong emotion. When conflict and emotion arise, these participants react in one of two ways—or in both ways at different times. They drop out of the interaction when things get too "hot," or they try to stop what's happening. When confrontation takes place—and I mean responsible confrontation—they engage in what has been called "Red-Crossing." Like Red Cross volunteers, they rush to mend the wounds of the person being confronted; they do their best to soften or put an end to the conflict. I don't suggest that all conflict is good or that conflict should be stirred up for its own sake. However, research studies have shown that conflict, if faced reasonably, contributes to rather than detracts from the growth of the group. On the other hand, conflict that isn't allowed to surface at all will keep the members from developing their relationships with one another. For instance, if the instructor is acting in a self-centered way—directing all interactions and perhaps making the group members engage in a lot of anxiety-arousing exercises—but there has been a silent decision not to confront the instructor, then this unfaced conflict will get in the way of the work of the group. If conflict makes you feel uncomfortable, it's best to let the rest of the group members know. They can help you monitor your attempts to avoid conflict and can help make you feel more skilled and comfortable in dealing with reasonable conflict.

Dealing with conflict involves three skills that you've already learned:

- disclosing what's troubling you,
- reasonably confronting the other person or persons, and
- responding nondefensively to confrontation.

Conflict ordinarily involves some negotiation and the ability to work out some kind of compromise. You're not asked just to give in; that would be asking you to be nonassertive or passive. You *are* asked not to just demand your own way, for that would be self-centered and aggressive. You're asked to meet your own legitimate needs and wants in ways that respect the needs and wants of others—that is, to be assertive.

Hostility. Hostility is one of those strong emotions that many of us fear. There was a time in the development of human-relations-training groups when the expression of raw hostility toward others was seen as one of the most liberating interpersonal experiences in which a person could engage. Those days, hopefully, have passed; raw hostility is now seen by most as a form of aggression rather than as assertiveness. Being hostile can be a way of running away from the work of the group. Some people use hostility to accomplish goals that might be accomplished more effectively in other ways:

• Hostility can be a way in which a person tries to express his or her individuality or to show his or her strength in the group. It can be a way of saying "I'm my own person" or "I'm a strong person." There are better ways of expressing individuality and strength—for instance, by being assertive in the ways that I've been describing.

291

• Hostility is a common reaction of people who feel threatened. Often the best defense is a good offense, and so people lash out before they can be attacked themselves. It's far more constructive to deal openly in the group with whatever may make you feel threatened. For example, you may feel threatened by the depth or intensity of others' self-disclosure, or you may feel threatened by the fact that people feel free to ask one another to live up to the provisions of the contract. Whatever the threat, see whether you can respond to it openly.

• I've known people who actually plan their hostility. For instance, they feel that they aren't getting through to another person, and so they unleash a blast of hostility to shake the person up. Or they use hostility as a dynamite technique to stir up some action during a boring session. Such techniques are indirect and manipulative, however. They imply a lack of respect for other people. If you want to stir things up, do so directly, responsibly, and in keeping with the provisions of the contract.

• Some people see hostility as a way of achieving intimacy. At times two people do tend to draw closer together after they've stormed at each other. Perhaps they find a more direct road to intimacy either too slow or too difficult; so they use hostility as a shortcut. This method might be all right if both people feel the same way. However, if you use hostility in this way with someone who doesn't feel the same way, you're merely being aggressive. Intimacy can be pursued through the various forms of immediacy I've already described.

Some people use confrontation as a way of expressing their hostility. They don't use confrontation as a way of inviting the other person to examine his or her behavior; nor do they see confrontation as a mode of interpersonal involvement. Rather, they pepper their adversary with a lot of questions and generally try to build a case against him or her. They come across as prosecution lawyers rather than as people trying to build relationships.

Another way of expressing hostility is to try to control and manipulate others. Often this is done through various forms of interpersonal game playing:

- I don't like you, so I become passive and make you come to me and try to get me to relate to you.
- I'm afraid of you, and so I become your "friend" and sweet talk you so that you can't hurt me. I try to seduce you to get closer to me.

- I'm annoyed at you, so I try to get others to confront you and suspect your motives.
- I see you as a more effective communicator than I am, so I put you into the role of "group leader" and then point out how you're failing to lead.

Hostility as described here is a form of flight, for it is doing something indirectly that would be better done more directly. Therefore, any use of hostility in the group should be examined—not because hostility is evil in itself, and not because a certain amount of hostility isn't normal in human relationships, but because hostility may really be a cover-up for something else.

Exercise 48
Flight: An Exercise in Self-Confrontation

We've so far reviewed some of the ways in which individual group members can run away from the work of the group. This exercise will help you identify the ways in which you personally might be tempted to flee.

Directions

a. Review the ways of fleeing described so far in this chapter. Make a list of the ones that pertain in some way to you.
b. Add any kinds of flight you use that aren't included in this chapter.
c. Check the one or two that especially interfere with your full participation within the group.
d. Share briefly and concretely with the whole group the one or two principal ways in which you resist the work of the group.
e. Indicate how you'd like to change this behavior. Be as concrete as possible.
f. Indicate how you'd like the other members of the group to help you change your behavior.

The Group in Flight

When a number of group members are in flight together, the group is faced with a more serious situation. Individuals in flight can be confronted relatively easily, but when an individual suggests that the group as a whole is running away from its work, he or she can much more easily be ignored or even punished. It often takes greater assertiveness or courage to confront the group than to confront an individual, but perhaps it's also more important to do so, since everyone is affected by the group in flight.

Confronting the group rather than just a single individual in the group is a separate skill. Like the other communication skills, it has three parts.

1. *Awareness.* You must first be aware of what's happening in the group. If you begin to feel uneasy in the group, ask yourself what's making you feel that way. For instance, you may discover that three of the seven group members have said nothing in the last 20 minutes. You begin asking yourself what has caused these three members to withdraw. If one person falls silent, it is, in a sense, an individual problem. However, if three members fall silent, it's certainly a group problem. If you *listen to the group,* you'll soon become aware that behaviors are taking place that make the group less productive.

2. *Communication-know-how.* A confrontation of the group should follow the general form for confrontation:

- *Describe rather than judge.* Your confrontation should be principally a description of the behavior or behaviors you see as unproductive. As with one-to-one confrontation, avoid judging, labeling, name-calling, and the like.

  > *You:* "I'm feeling uneasy right now. Four of us haven't said anything in the last 20 minutes or so. It's almost as if the other three were having a private conversation and we were just spectators. You three are really enthusiastically into your conversation. I'm not sure what's happening with the rest of us."

  Here you don't blame anybody for anything, but you do describe what's happening.

- *Include yourself.* You should indicate that you feel responsible for your own behavior and for whatever it contributes to the problem in the group. Don't stand apart from the group and become its critic. Remain mutual by including yourself.

> *You:* "In one way I feel excluded from the conversation, but, if I think more carefully, I'm probably excluding myself. You're talking about sexuality, and that's a very sensitive topic for me."

- *Invite the others to examine what's happening.* You have the right to invite the others to examine what's happening in the group. After all, it's as much your group as it is theirs.

> *You:* "I'd like to find out how the rest of you feel. I'd particularly like to find out why the rest of you have become silent like me."

This invitation presents the situation as a group responsibility.

Assertiveness. As we've seen, all communication skills demand a certain degree of courage or assertiveness, a willingness to take a risk. Since the risk may well be higher when one person confronts the group, it's helpful if, at the very beginning of the group, the members discuss the possibility of group confrontation. A discussion at the beginning reinforces the fact that it's quite *legitimate* to ask the group as a whole to give an account of itself. The viewpoint of the one who confronts the group may not be shared by everyone, but he or she should have the support needed to air his or her views.

Forms of Group Flight

Here are a few of the ways in which the group as a whole can run away from the work of the group.

Analyzing. Prolonged analysis of past interactions is a form of group flight as well as individual flight. If this kind of behavior isn't confronted early in the group, it becomes one of those silent decisions that make the group less productive than it can be. Time after time groups follow an ounce of interaction with a pound of analysis. Remember that a sign of flight through analysis is the almost exclusive use of the past tense by the group members.

> *You:* "I've noticed that in the past few minutes we've been using the past tense and discussing the past quite a bit. Ted, you and I have been talking about how we felt about each other last week. I'd like to get down to how we feel about each other now that we've had a week to think about it."

Analyzing isn't examining interpersonal style or establishing and developing relationships. Analyzing isn't attempting to change behavior. Analyzing is really a way of taking time out. It's usually too vague and heady. It has no impact on the here and now. The point is not that the past should never be mentioned but that it should be quickly connected to the present.

> *You:* "I was confused last week when you confronted me about talking too much. I thought I was being assertive. But, now that I've had more time to think about it, I see that you've got a point. I try to bury people with words. I keep people at a distance because I've got the gift of gab. I feel I could do it again right now."

Here you connect the past quickly to the present and make a real attempt to examine interpersonal style as this style affects the group.

Some analysis and some processing of what's taking place in the group may occasionally be worthwhile, but it's very often over-done. It would be far different if the group were involved in fast and furious interchanges for quite a while and then somebody said "We've been very intense for almost an hour and a half. Could we perhaps stop a moment and take a look at what we've been doing?" This is a long way from 15 minutes of significant interaction followed by 45 minutes of analysis or processing.

Focusing on one. In many groups, the members fall into the pattern of taking turns being the center of the group's attention. One individual talks about his or her problems and then gets feedback from everyone else. You might hear a person say before a group meeting "I bet I'm going to be in the hot seat today." I see this pattern as a form of group flight for a number of reasons.

First, giving each member "time" to deal with his or her "problems" is more like therapy or counseling. The person who is on the "hot seat" for an extended period of time almost inevitably becomes, at least for the time being, a client or patient. The principal

296

characteristic of a human-relations-training group, *mutuality*, disappears.

Second, when this takes place, often enough a number of group members say nothing to the individual. Not only is it counseling, but there are a client, a couple of counselors, and then a few observers, the ones who have nothing to say to the client. No way of proceeding that significantly reduces the involvement of a number of members or gives them an excuse to reduce their own involvement fits into the goals of the human-relations-training group.

Third, focusing on one person for a great part of any group meeting involves another self-defeating process: an attempt to finish with a person or with some particular issue at one sitting. No person or issue can be finished with at one sitting. For instance, if you have trouble being assertive (you don't initiate conversations, you fail to join ongoing conversations very often, and you don't confront because you feel too anxious), your lack of assertiveness won't be "cured" merely because the group devotes an entire meeting to your problem. It would be much more useful for you to get some initial feedback on your nonassertive style together with some discussion on how you might go about changing that style. Then some gradual attempts to change from meeting to meeting are called for. If you get too much feedback from everybody at one meeting, you'll be *overloaded*. You might well begin to feel "I can't handle all of this. I might as well forget about it." Consistent effort and pacing are important elements in your attempts to change your interpersonal behavior.

Fourth, focusing on one person defeats one of the group goals—establishing and developing relationships. In trying to establish and develop relationships with one another, the group members should make feedback a continuous process, just as relationship building is a continuous process. If I'm aggressive today and passive tomorrow, I need feedback on my aggressiveness today and on my passivity tomorrow. It doesn't help me if you save up your perceptions until next month and then tell me "Sometimes you're aggressive, and sometimes you're passive."

There are times when giving an individual member extended time in the group to deal with a certain issue *is* called for, but I prefer groups that don't make a *habit* of dealing with individual members in an extended way. Some people get the idea that the only way to support a person is to give him or her all of the time he or she wants in the group. I see that as phony support. Support, like feedback, should be continuous. One of the most effective forms of support is

297

mutual self-sharing, and that's impossible to achieve in a focusing-on-one-at-a-time group.

Doing the same thing all the time. Sometimes groups become too dependent on structure. In our lab experience so far, we've used a great deal of structure. It's time to put most of the structure aside and to use both individual and group-specific skills more spontaneously. Groups that have become overly dependent on structure sometimes manufacture their own. Very often this new structure is the kind that keeps the group safe. In this kind of group, one meeting seems to be a carbon copy of the next. The members are often very good at both individual and group-specific skills, but they lack spontaneity and intensity. If you were to videotape five successive meetings in such a group, it would be difficult for anyone watching the videotapes to determine which meeting was first and which fifth. They all look alike; there's no deepening of relationships; there's no increase in the intensity of interactions. In terms of skills, the group is technically good, but it's going nowhere.

Group members can avoid this kind of flight if they come to each meeting with solid agendas that involve some reasonable risk taking. But the risk taking can't be left up to just a few. Mutuality demands that all the members participate, to one degree or another, in the risk-taking process. Risk taking can include trying out the kinds of interaction that a given person finds particularly difficult. For instance:

- Penny has become a specialist in communicating basic understanding. She risks herself by including more self-disclosure and confrontation in her agenda.
- Larry always begins the group interaction and takes the most risks in self-disclosure. He changes by waiting for others to disclose themselves first and by providing more basic understanding for others when they do risk themselves. This kind of behavior is a risk for Larry, because his picture of himself and his good feelings about himself revolve around his being the most active member in the group.
- Jason has dealt more intimately only with three of the other six members of the group. He has been comfortable that way. He risks himself by dealing with "you-me" issues with the other three members.

298

There's one ritual that's fairly common to human-relations-training groups. Almost every group has a relatively quiet member. Often, when there is a lull in the group, the members (silently) decide that it's time to deal with the silent member again. They turn their attention to this person for a while, tell him or her about his or her silence, and receive the same kinds of replies from the silent member. After this ritual is finished, they go about their business once more. Nothing really changes, and the whole process is simply an indication that the group can get into certain ruts.

The high-level group asks itself from time to time whether it has fallen into these kinds of ruts in order to keep the interaction safe.

Moving at the pace of the slowest member. There are usually quite noticeable individual differences among members of any human-relations-training group. Some of us learn communication skills faster than others. Some of us develop the ability to trust and to risk ourselves faster than others. The problem here is that groups have the temptation to move at the pace of the slowest member. It can even be worse if the slowest member is slow because he or she is indifferent to the goals of the group. Getting people to commit themselves to a contract *before* the group begins is one way of keeping the unmotivated person out of the group. However, if an unmotivated person does get into the group, I think that it's better to ignore such a person than to spend a great deal of the energy of the group trying to motivate him or her. This is both unfair and a form of flight from the real work of the group.

The question of the person who is merely slower than the others is a different one. You can be sensitive to such a person without giving up your own right to a healthy, reasonably intense group interaction. There's a middle ground between ignoring a slower member and making him or her the center of the group's attention. Meeting your own needs while simply ignoring the slower member is a form of aggressive behavior. Setting your own wants and needs aside and spending most of your time trying to help the slower member is at the other end of the scale—nonassertive behavior. An assertive group member neither ignores the slower member nor sets the pace of the group at a level that will make such a member feel uncomfortable.

Engaging in serious conversations that aren't goal directed. Sometimes groups run away from the real work of a human-relations-training lab by having very serious and worthwhile conversations that have no direct relation to the goals of the group. I once sat in on a training group that was having a very high-level discussion of some of the more important social questions that face the United States today. Had a stranger come into the group without knowing its purpose, he or she would have assumed that it was a social-action discussion group. Such serious conversations are important, but not as part of a human-relations-training laboratory.

Some participants are willing to talk about serious problems that are affecting their lives *outside* the group but are much less willing to talk about their relationships to other group members. Some younger people, for example, can talk at great length about their relationships with their parents or about people outside the group with whom they want to have intimate relationships. Admittedly, these are serious concerns, but they're out of place if they focus the attention of the group outside. It's a far different matter if people talk about relationships they have outside the group but relate what they say to the group members and to their general interpersonal style.

Non-goal-directed, serious conversations aren't easy to challenge, because they seem so good in themselves. Therefore, it's too easy for a group to make a silent decision not to challenge any conversation that is about some serious topic, even if the topic isn't related to this group or to the general goals of the group. If you find a willing and understanding audience in your fellow group members, you may well be tempted to talk about *all* the things you find difficult to talk about outside. However, one of the goals of the group is to teach you how to establish these same kinds of relationships outside the group. The training group is meant to help you be more effective interpersonally in your everyday life; it's not a substitute for real life. You may want to start up a different group outside, in which you can talk about all sorts of serious life issues.

Avoiding intensity. Although not every group meeting should be a draining experience for every participant, group interaction *is* work, and you should be reasonably tired at the end of a group meeting. Some groups spend more energy making sure that nothing happens than they would if they worked for strong group interaction. I've talked about silent decisions that rob groups of their energy and vitality. An alternative to making silent decisions is to do just the opposite—to explicitly make "rules" that help give the group more energy and vitality. Some of these rules might concern:

300

- *what can be talked about.* "We can talk about *anything* that helps us achieve our goals of examining interpersonal style, establishing and developing relationships, and changing the behaviors that we're dissatisfied with."
- *the ways in which the group proceeds with its conversations.* "It's all right to interrupt somebody in order to join an ongoing conversation and contribute to it. Interruptions are necessary for mutuality."
- *intensity of interactions.* "It's all right to express strong emotion, such as anger, if it is done assertively rather than aggressively. Reasonable expression of strong emotion that can be worked out in the group isn't the same as dumping."
- *the contract.* "Since we've all agreed to the contract, no one has to apologize about putting it into practice or asking that others put it into practice."
- *interpersonal style.* "No one can violate the contract in any regular way by merely claiming that what he or she is doing is simply part of his or her interpersonal style. For instance, no one can take flight through humor and claim 'I'm just naturally a funny person. Humor is important to me.'"
- *group goals.* "We should review our behavior from time to time in order to see whether it fits in with the goals of the group. For instance, one goal is to form ourselves into a learning community. We need to step back from time to time to see whether this is happening."
- *leadership.* "Leadership isn't just a function of the trainer or instructor. This group will go well to the degree that all of us participate in the leadership function of the group. Whenever anyone in the group is doing something that's in agreement with our contract, he or she is showing leadership."

We've now reviewed some of the ways in which the group as a whole can flee the contract and the work it calls for, together with some suggestions for challenging these modes of flight. Because we're all human, all the group members are going to be tempted from time to time to run away from the work of the group. The point is that the group members should challenge themselves and one another when this happens.

Exercises in Dealing with Group Flight

Exercise 49
*Challenging the Unproductive "Rules" of
the Group*

Directions

Write out a list of the rules that have been made silently and that rob the group of energy, vitality, and intensity. Of course, if you think that your group hasn't made any such rules, don't manufacture some just to do this exercise.

Examples

- "I think we've made a rule that it isn't necessary any more to provide much basic understanding. People are being asked to assume that the others understand them."
- "It strikes me that we've made a rule that we must treat everyone in the group the same way. We can't show that we like one person more than another. If we do like someone particularly, we can show it only outside the group."
- "There seems to be a rule that says that we start each group meeting entirely fresh. We don't take up where we left off at the last meeting. Therefore, we seem to end up with a lot of unfinished business in the group. I'd like to begin some meetings by asking what unfinished business we have."

Now share your findings in the way determined by your instructor.

- See how many members have come up with the same rules.
- Determine which rules are doing the most harm in the group.
- With your fellow group members, determine how you'll go about changing these unproductive rules.

Exercise 50
The Emotions We Avoid Expressing

This exercise will help the members of the group to identify emotions that aren't being expressed by the members of the group and to expose the rules that the group has invented to keep the emotional climate safe.

Directions

This exercise can be done in the full group. Each member tries to think of some emotion that he or she feels "shouldn't" be expressed in the group and the reasons why they shouldn't be expressed.

Examples

- *Member A:* "I shouldn't express curiosity about what others haven't disclosed about themselves, because this is butting into other people's business."
- *Member B:* "I shouldn't get angry at anyone, because the person I get angry at and perhaps others will get angry back, and I can be hurt."
- *Member C:* "I shouldn't express affection in here, because if I do I run the risk of being rejected."

These "rules" can be written down on a blackboard or on a large sheet of paper for all the members to see. The members can discuss how these rules are affecting interactions in the group. It's usually easier to deal with both personal and silent-majority rules once they've been brought out into the open.

The Difference between Flight and Adequate Defenses

Psychological safety is an important issue in human-relations-training labs. For instance, if one of the members begins revealing inappropriately all the secrets from his past, and if he experiences tremendous anxiety in doing so, his self-disclosure could be quite harmful. He might not be able to handle all that anxiety. It might begin to interfere with his work, his recreation, and his relationships. Most psychologists suggest that it's necessary to defend ourselves against destructive anxiety. We all need to be reasonably careful about our psychological safety. Some people, however, defend themselves *too* well. They don't take any psychological risks, because they don't want to experience any anxiety. For instance, there are people who even in marriage engage in practically no deep sharing of themselves. Such people, by defending themselves too well, lose out on a lot of life. They remain psychologically safe, but they never experience the rewards that come from risking intimate relationships.

What I am suggesting in this book, therefore, is that you lower your psychological defenses in the lab and that you take more risks than you might in your everyday relationships, not for the sake of taking risks but for the sake of exploring further interpersonal possibilities. In the lab you can stretch yourself interpersonally, experimenting with new or untried forms of interpersonal behavior to see what they might add in a positive way to your interpersonal style. The structure of the lab and the support you receive from the other members of your learning community provide the safety you need to lower your ordinary, day-to-day defenses in reasonable ways. You can still maintain adequate defenses so that you're not overwhelmed. Take self-disclosure, for example. Most of us could afford to experiment with increased self-sharing without being overwhelmed by anxiety. However, if we were asked to share our deepest secrets and, generally, to engage in a kind of psychological nudity, this would understandably provoke great anxiety. In one case we lower our defenses; in the other we throw our defenses to the wind. You'll probably find that for you, as for most people, the more common danger is to fail to lower your defenses enough to allow the laboratory experience to have an impact on you. The person who has poor psychological defenses to begin with usually avoids laboratory experiences like this one. Too many lab participants end the lab by saying that they wish they'd done more or that others had made more demands of them. Participants commonly say at the end of a lab experience that the other members of the group didn't confront them enough. If you feel that you're not being challenged enough, it may be that you're giving others subtle cues that say "Don't go too far with me."

Some people leave a group saying that it didn't do much for them. Others leave the same group saying that it was the most growthful interpersonal experience they'd ever had. The truth seems to be that labs affect people to the degree that people give themselves to the lab experience and work for the lab goals. There is no magic. There's only hard work.

As I've observed, the need for structure in the group lessens as people learn skills—both individual and group-specific—and use them to pursue the goals of the group. Consequently, the need for formally structured exercises also lessens. However, even at this stage exercises can help break up logjams and help the members of a group focus more carefully on some issue. Exercises, then, shouldn't be introduced just to stir something up; if they're related to the goals of the group, they may help. An excessive use of exercises at this stage of the group is either uncalled for or else indicates that the motivation and initiative of the group members are low. If this is the case, the issue of motivation should be examined by the members of the group instead of being covered up by the use of exercises.

Exercise 51
Unanswered Questions: Discovering Hidden
Immediacy Issues

The principal goal of this exercise is to help the group to get at immediacy issues that don't surface in ordinary group interaction. Remember that by *immediacy issues* I mean "What's going on between you and me?" A secondary goal is to enable you to learn by experience something you've already read about: how questions are often really statements and not questions at all.

Directions

Let's say that your group has six members—A, B, C, D, E, and F. Any member begins by asking another member a question that relates in some way to the group or to the goals of the group. For instance, A asks D "Do you like being in this group?" (This could be an example of a question being really a statement; that is, A might *mean* "There are ways in which I see you being uncomfortable in this group.")

Next, the member asked (in this case D) *does not answer the question.* Instead, he or she asks *someone else* a question. For instance, D asks F "Why are you hesitant to confront me?"

No one may ask a question *right back.* That is, when A asks D a question, then D may ask question of anyone *except* A. Later on, however, if C asks D a question, then D can ask a question of A. This prevents two people from having a "question debate" that would exclude the other members of the group.

305

The questions should be related to the goals of the group and go on for five or ten minutes. Usually, so many issues come up that it's hard to keep track of them if the questioning goes on too long. Each person should keep a mental note of the issues he or she wants to pursue after the questioning stops. It's important to *use* the information that surfaces in the course of this exercise. Too often, groups bring up excellent issues but then fail to discuss them after the questioning stops.

Finally, this exercise helps the members of the group see that often many significant issues remained *buried.* The exercise can stimulate a lot of thinking about such buried issues. However, once participants come to realize that they tend to sit on certain "you-me" or group-relevant issues, they need to assert themselves and bring up these issues without the need of an artificial exercise. If this exercise were to be used often, the members would surrender their initiative to an artificial structure.

Exercise 52
Trust and Immediacy: The "Undisclosed Secret"

If the group is to take reasonable risks, trust must continually deepen. The purpose of this exercise is to help you discover snags in the trust-building process.

Directions

Think of some secret or something else that *you don't want to disclose* to the members of your group (it could be something you would consider inappropriate to disclose). The less you would like to disclose it the better. In choosing your topic, *remember that you'll never actually disclose it to anyone.*

Next, *in your imagination,* see yourself trying to disclose your secret *privately and separately* to each of the other members of your group. In your mind's eye, see yourself disclosing (or not being able to disclose) your secret to each. Try to experience how you would feel with each person

After you've completed this task, share what you *imagined* (*not* your secret) and how you felt with your fellow group members in the way determined by your instructor. For instance, you might find yourself saying such things as:

- "Tom, it was easy to tell you, because I knew that you would be understanding. Eventually, I think that you'd challenge me on this issue, but not before understanding me."
- "Adele, I just couldn't tell you. I was too embarrassed. I guess I felt that it would be too embarrassing for you. I said to myself that it would make you too anxious, that you couldn't handle it. This may be unfair to you, but maybe that's how I relate to you in the group."

Again, this exercise might stimulate you and the other members of the group to face issues of trust more directly, but it's not meant to be a substitute for your doing so on your own initiative.

Exercise 53
Immediacy: The "Person Who———Me Most"
and/or "What ——— Me Most about You"

This exercise presents a way of stimulating your thinking and feeling about both confrontation and immediacy issues.

Directions

First, a suitable verb is selected by the instructor or group members to complete the sentences in the title of this exercise. For instance, "The person who *puzzles* me most in this group is . . . ." Or, for the second sentence, "What *stimulates* me most about you is . . . ." Obviously, a number of different verbs may be chosen—"helps," "challenges," "annoys," "puts me off," and so forth. What different verbs can you suggest that would be related to the goals of the group?

Decide whether you want to use the first or the second sentence. The first sentence has you single out one person from the group. For instance, "The person who puzzles me most in the group is you, Tom." The second sentence is used to address each of your fellow group members. For instance, "What challenges me most about you, Annette, is that you can be honest and still caring in each of your relationships here. You don't *like* everyone the same, but you certainly *respect* everyone."

307

Whichever sentence you use, take time to think of the *concrete* reasons for the way in which you complete it. In the case of the first sentence, what in this person's behavior moves you to choose him or her?

> "What puzzles me about you, Tom, is how you differ from week to week in your participation. One week you're very enthusiastic and initiating. The next week you can be very quiet. Sometimes it's like having two different people here in the group."

In the second sentence, what precise behaviors challenge (annoy, stimulate, help) you?

> "Annette, you seem to find time to initiate conversations with everyone here. You attend, you respond—all of this comes across as genuine."

If you use the first sentence, don't gang up on any one particular member. Remember that, since this exercise may involve both confrontation and immediacy, follow the rules of responsible confrontation in what you have to say.

Leadership and Five Different Types of Group Participation

As you may have noticed, I've said little or nothing so far about leadership. In a sense, however, this entire book is about leadership—or rather about the kinds of behaviors that make for good leadership. I don't want to stress the person of the instructor, even though he or she has an important leadership role in the group. The main thing is that you and your fellow group members initiate skilled interactions with one another. When this happens, everyone is participating in the leadership function of the group. Carkhuff (1974) talks about five different types of participants in a group or in a program. These types can help us understand leadership and how all the group members can be leaders in the group. The five types are the Detractor, the Observer, the Participant, the Contributor, and the

Leader. Let's take a look at each type in the context of a human-relations-training group.

The Detractor. The Detractor not only fails to work for the goals of the group but is an obstacle for others. If the Detractor participates in the group, he or she does so in a destructive way. Let's take a look at some typical varieties of Detractors.

- *A person needing counseling.* Some people may need counseling before they're ready for the kind of human-relations-training group described in this book. They have such serious problems in relating to others that they tend to focus all of the group's attention on themselves. It would be better if they pursued another kind of group experience first—a counseling or therapy group. In that kind of group they could explore their problems, come to a new understanding of themselves, and take action to change their self-destructive behaviors.
- *The "intellectual."* Some people find it very difficult to act without first discussing things intellectually at great length. These people are really looking for a discussion group rather than for a training group. They question everything, have theories of their own, quote books they've read, and tend to give intellectual interpretations to the behavior of the group members.
- *The person who doesn't really accept the contract.* Some people say "yes" to the contract in order to get into the group (they want course credit, they think it might be an easy course, their friends are doing it, and so forth). Once in the group, they accept parts of the contract and reject others. For instance, they accept individual skills training but aren't interested in making the group a learning community. They don't disclose very much about themselves and don't respond to those who do.

The kinds of flight described in this chapter are really Detractor behaviors. What other kinds of Detractor behavior could you list? Usually no one is an out-and-out Detractor. The pure Detractor type usually doesn't get into such groups in the first place. However, some of us can fall into Detractor types of behavior from time to time. If you do this, you can expect to be challenged.

The Observer. The person who doesn't reach out actively to initiate conversations with other group members and who, when contacted by others, responds poorly is an Observer. Like the Detractor, an Observer is a burden on the group. It would be better to screen out potential Observers before the group begins, but it's sometimes hard to spot an Observer in the early stages of the training program. Since structure is prominent in the early part of the group—training in individual skills—the Observer often isn't noticed. It's impossible not to participate in a highly structured skills-training program. However, later in the life of the group, when the structure is less important and each individual member has to take more initiative, the Observer is seen clearly. The idea of a relatively silent but harmless group member who is "learning a lot just by observing" is nonsense. Observers cannot remain Observers very long in a group without becoming Detractors.

Anyone who shows signs of becoming an Observer can be helped to examine his or her behavior *early* in the life of the group. If the other members don't make legitimate demands on the potential Observer right from the beginning, they're being unkind both to themselves and to the potential Observer. It goes without saying that the fears and difficulties of the potential Observer need to be listened to and understood, but it also goes without saying that he or she needs to be helped to face these fears early in the life of the group. One way of doing this has already been suggested. *All* the members of the group can talk about their fears of participating fully. In this way, the potential Observer will see that he or she isn't alone in having fears. He or she will be able to put his or her fears in the wider context of the group and won't end up becoming the group patient. The potential Observer may still need more support and encouragement than other members do to participate adequately, but that's understandable and doesn't stand in the way of the goals of the group. However, once a person becomes identified with the role of Observer, it seems almost impossible to get rid of the role.

The Participant. Whereas Detractors and Observers interfere with the work of the group, Participants do not. Participants respond in some positive way to goals and to the structure of the group. For instance, Participants cooperate actively with the skills-training parts of the lab. When it comes to group interaction, Participants tend to respond when others initiate rather than to initiate their own interactions. Participants respond, however, not reluctantly but willingly and display a good deal of cooperation in responding. They are *active* responders. At worst, a Participant is overly dependent on other people's initiatives. Therefore, if a group has too many members who are merely Participants, it runs the risk of being a relatively dull group.

As with the Detractor, perhaps there are few people in training groups who fit the description of the Participant completely. However, there is a danger of having too many of the group members assume the Participant role too much of the time. Being an initiator means both work and risk. It's easy for a number of group members to slip into the comfort of the Participant role. The Participant seems to say "I'm here all right. I'll certainly cooperate. But don't expect too much of me in the way of getting things started." If you slip into the role of a Participant in the group, what you gain in terms of comfort you'll probably lose in terms of interpersonal growth.

The Contributor. Ideally, your goal in the group is to become a Contributor. The mark of a Contributor is initiative or assertiveness. The Contributor is a self-starter. The Contributor doesn't wait to be contacted by others but rather actively initiates conversations. He or she knows that it takes a great deal of assertiveness and work to establish and develop relationships with each of the other members of the group, so he or she gets to work immediately. *Contributors make groups go.* Contributors develop at least minimally adequate levels of both individual and group-specific skills and also the assertiveness and courage to put these skills to good use. In a group in which all the members are Contributors, each has to fight for time to get his or her agenda taken care of. The Contributor keeps a good log and develops good agendas, but, more than that, he or she puts agendas into practice. If your group has frequent low or dull periods, there are probably too few Contributors or too many who let themselves fall into the Participant or even the Observer roles.

The Contributor provides high levels of human nourishment in the group by responding with basic, accurate understanding and by showing active, working respect for fellow group members. The Contributor is self-sharing in appropriate but risk-taking ways. He or she isn't afraid to challenge either self or others. If the Contributor isn't getting feedback on interpersonal style, he or she asks for it, displaying assertiveness rather than passivity or aggression. The Contributor joins ongoing conversations without apologizing and with-

311

out disrupting them. Mutuality is important to Contributors, and they make good, functional members of the learning community.

The Leader. This seems the best place to describe briefly the function of the trainer or instructor in the human-relations-training group, for the instructor is expected to be a Leader. In the first part of the group, the instructor exercises leadership in a number of ways. He or she explains the theory underlying the skills, demonstrates the skills themselves, trains the members of the group in the skills, evaluates the performance of the participants, and teaches the participants how to give one another feedback on performance. The instructor both supports and challenges members to acquire these skills.

In the group experience itself, the instructor has several functions. First of all, he or she is a Contributor, doing all that the role of Contributor implies and using communication skills spontaneously, assertively, and effectively. In my opinion, if the instructor isn't a Contributor, he or she cannot provide other leadership functions.

Second, the instructor encourages and challenges others to become Contributors. In a way, he or she works himself or herself out of a job. As more and more members become Contributors, less and less structure or direction is needed from the instructor. Sometimes group members idealize the instructor as *the* Contributor, giving themselves an excuse to remain in the Participant role. Since it's quite easy both to idealize and to fear the instructor, you might well take up this "you-me" issue with him or her, early in the life of the group. If you're hesitant to talk to the instructor, or if you find yourself waiting for him or her to get things going, these may be signs that your expectations of the instructor are unrealistic.

Third, the instructor is in large part responsible for giving direction to the group experience. For instance, he or she introduces skills and training methodology, assigns exercises, and establishes timetables for the introduction of new skills. He or she tries to keep the group moving at a pace that forces the group members to stretch a little. The instructor doesn't apologize for giving direction to the group. He or she realizes that the members have bought into the contract and that, consequently, he or she doesn't need to defend the process or to try to coax the members of the group into trying skills or exercises. Vince Lombardi, the late coach of the Green Bay Packers football team, was once asked the secret of his team's success. He replied that his team was excellent in basic skills—they practiced until they could block and tackle flawlessly. I don't imagine that he and his assistant coaches asked for volunteers to tackle the dummy.

They didn't apologize for their training methodology. The same holds true for the instructor or trainer in the group. While being understanding, caring, and supportive, he or she is also challenging and demanding. Only trainers that are Contributors can make demands of others.

Changing Your Interpersonal Behavior

This chapter suggests how to go about applying what you learn in the group to your everyday life. The basic process of changing human behavior is described and then applied to the specific task of changing your interpersonal style.

Once you've explored your interpersonal style by revealing yourself, by watching yourself interact with your fellow group members, and by getting feedback in the group, you may discover some features of your style that you'd like to change—behaviors that you'd like to acquire, lose, or modify. To help you in this task, let's look at what is necessary to change behavior.

If you want to change your behavior—in the present case, your interpersonal behavior—there are basically three things for you to do.

1. *Explore.* Spend time exploring your present interpersonal behavior. Learn what you like about it and what you don't like about it. Identify behaviors that you now engage in that you'd like to get rid of (for instance, responding aggressively and angrily whenever anyone does anything to annoy you). Also identify behaviors that aren't part of your present interpersonal style but that you'd like to adopt (for instance, responding frequently to others with basic, accurate understanding). You won't be able to change your behavior unless you first find out what you're actually doing or not doing. Furthermore, you should be as concrete as possible in identifying the behaviors you want to change.

You've been spending a great deal of your time in this interpersonal lab exploring, as concretely as possible, the behaviors that make up your interpersonal style. By now you should have a good idea of what behaviors you'd like to get rid of and what behaviors you'd like to acquire.

2. *Get new perspectives.* Since most of us have blind spots in trying to get accurate pictures of ourselves, we need the help of others (friends, teachers, acquaintances, relatives) to see ourselves as others see us. When others give us feedback on our interpersonal styles, we get a bigger, more concrete, more detailed, and more useful picture of ourselves. When others challenge and confront us, we begin to see other possibilities. For example, I may come to see that, if I make my humor less biting and sarcastic, it will become a plus instead of a minus in my interpersonal style. I may not have thought of that before, because I didn't see myself as others saw me, and I hadn't even thought of the possibility of humor without sarcasm.

Through the communication of deeper understanding, through confrontation, and through immediacy exchanges in the lab, you've been getting the new perspectives of yourself that you need in order to change. You've begun to see not just what behaviors you might drop but what behaviors you can put in their place. Getting new perspectives means getting a picture of yourself as you would like to be in your interpersonal life.

3. *Act.* Once you see point A, or what you're like now, and point B, or what you'd like to be, the next step is to do something to get from point A to point B—that is, to come up with an *action program* to get you where you want to go. For example, suppose that you now see that you're very passive when you meet a new person. You wait for the other person to start the conversation, and even then you say very little. This is your point A. In the lab you get a picture of a different possibility. You see that it's possible to become a person who stops avoiding new people, who starts conversations with them, and who even feels comfortable in doing these things. This is point B. Now the question is: how do you get from point A to point B? The answer to this question will give you your action program. In this case, you see that part of your action program should be to learn such fundamental interpersonal skills as self-disclosure and the communication of basic understanding very well, so that they become second nature to you. You see that this will help you to gain the self-confidence you need to meet new people. You see that these skills can be part of a method of meeting new people creatively.

If exploring, getting new perspectives, and acting are the heart of any change-of-behavior program, then the interpersonal lab is already giving you most of what you need in order to change. Let's

316

review the lab from the point of view of a change-of-behavior program.

The Lab as a Change-of-Behavior Program

The following steps constitute a change-of-behavior program within the lab itself.

1. *Learning core interpersonal skills.* Learning the skills of self-presentation, responding, and challenging is a behavioral-change program in itself. Furthermore, these skills are the building blocks of further interpersonal change. Trying to change your interpersonal style without these skills would be self-defeating.

2. *Getting feedback on your skills.* As you learn these core skills, the group provides you with the feedback you need to find out how well you're learning them. With the help of this feedback, you can become technically good at the skills.

3. *Learning group-specific skills.* Learning how to use these skills with different kinds of people—the other members of your group—is also a behavioral-change program. For some people, the greatest change at this stage of the lab is learning how to be more *assertive* in interpersonal situations, especially in groups. For instance, when somebody confronts you for not using basic understanding very much, you ask him or her to join you in exploring this dimension of the interpersonal styles of both of you.

4. *Practicing interpersonal assertiveness.* As you learn both individual and group-specific skills, you begin to put aside some of the passivity of your interpersonal style. You're less passive because you can now do things, interpersonally, that you couldn't do before. Since you're less awkward in interpersonal situations, you're also less likely to be defensive, hostile, and aggressive. This increased ability to initiate interpersonal contact in caring, respectful ways is a middle ground between being passive and thus not getting what you legitimately want and need from others and being aggressive and thus overwhelming others in your pursuit of what you want from them.

5. *Discovering patterns.* By continually discussing your interpersonal style, and, even better, by living out your changing interpersonal style in establishing and developing relationships with your fellow group members, and by getting feedback on this style, you begin to discover that you have certain interpersonal *patterns*, or characteristic ways of behaving with others. For instance, you may discover that you use basic understanding naturally and easily in

317

one-to-one conversations but that you practically never use it in a group. Or you might discover that, when you're confronted in a group, you withdraw; that is, you say little in return, you don't argue, and you don't explore the content of the confrontation—you forget it.

6. *Getting feedback on patterns.* Once you begin noticing some of your interpersonal patterns, you check them out with other members of the group. Either they give you feedback spontaneously, or you ask for it. You try to clarify the differences between the ways you see yourself and the ways others see you. Notice that you use both the individual and the group-specific skills you've learned in order to do this clarifying.

7. *Recognizing payoffs.* If you like some of the patterns you see in yourself, realizing that they contribute to a healthy and satisfying interpersonal style, you rejoice. However, if you see patterns that you don't particularly like, patterns that turn others off or that make interpersonal living less enjoyable or more difficult, you begin asking yourself why you hold on to these patterns. For example, suppose you see that your usual approach to meeting someone new is to be passive, quiet, and even mousy. However, once you get to know the person a little and lose your fear of him or her, you become demanding, dominating, controlling, and hostile if you don't get your own way. This pattern turns other people off. You therefore ask yourself why you act this way. The assumption is that, if you keep doing something that doesn't seem to be very growthful, there's probably some other payoff in it for you. What are you getting out of this negative pattern of interpersonal behavior? In this case the payoff may be that you never have to work at real closeness or intimacy. You're either fearful and passive or hostile and dominating, neither of which involves much discipline or work on your part. Getting out of work might not sound like much of a payoff, but a closer look will show that many of us give up some of the nobler things in life because we prefer the comfort of doing nothing.

8. *Seeing different possibilities.* In the group, once you discover self-defeating interpersonal patterns of behavior, you use the resources of the group to explore other possibilities. For instance, if you see that you usually respond to confrontation either by saying nothing or by getting annoyed, you can begin asking yourself what other possible ways there are to respond to legitimate confrontation. One way is to try to understand more fully what the person confronting you is trying to say. Another is to check with other people to see whether they feel the same way as the person doing the confronting. A third is to ask others for feedback on your interpersonal style *before* they confront you. A fourth is to confront yourself before others con-

front you and then to check out your self-confrontation with others. All of these possibilities are more promising for interpersonal growth than avoiding or resisting confrontation.

9. *Experimenting with new behavior.* The group experience encourages you to try different, more rewarding and useful, patterns of interpersonal behavior. For example, instead of getting hurt and pouting when you're challenged in some way, you can (a) communicate accurate understanding to the one who confronts you, (b) ask both that person and the others in the group to help you explore the behavior that is being confronted, and (c) try to come up with new possibilities.

10. *Evaluating yourself and getting feedback.* Once you begin to experiment with new patterns of interpersonal behavior in the group, you can both evaluate yourself and get feedback from others on how you're doing. You can tell others how you feel about your new behaviors and ask them how they feel about and react to this new you. If a new pattern feels good to you, and if the reactions you get from others are positive, then you can consider the possibility of using this pattern of behavior in your everyday life.

11. *Transferring what you learn.* Using new behavior patterns learned in the group in your life outside the group is called *transfer of learning.* Even at the beginning of the training program there is the possibility of transfer; that is, the individual skills you learn can be used almost immediately outside the group. As I've already noted, you'll probably find that increasing your use of basic understanding outside the group can do much to make for smoother interpersonal exchanges.

The group, then, is in and of itself a program of behavioral change. However, in order to make these changes permanent and transfer them to your everyday life, you should become familiar with certain *principles of change* that underlie any change program. If these principles can seep into your bones, the process of change won't seem quite so mysterious or difficult.

Principles Involved in the Change of Human Behavior

Motivation

You'll never change unless you want to. *You* must want new patterns of interpersonal behavior because of the rewards that you see in them. If you want to change merely because *others* are dissatisfied with you or want more from you, then you shouldn't expect too much from a program of change. Also, the good qualities you see in others may help you *begin* the motivation process, but a desire to have these qualities for yourself still has to get into *your* bones. For instance, you may admire the assertiveness that you see in others and become dissatisfied with your own passivity, but any effort toward change of behavior won't work until you actively want assertiveness as part of your own interpersonal style.

Being Concrete

In order to change your behavior, you need to know precisely what interpersonal behaviors you would like to acquire, lose, or modify. You can start with a general statement: "I want to be less passive in my dealings with others; I want to be more assertive." The next step is to spell out in behavioral terms exactly what you mean by "less passive" and "more assertive." For instance you might say:

- "I don't want to remain silent for long periods of time when I'm with a group of people."
- "If I feel emotions when I'm with others, I want to be able to express them. For instance, if I'm angry I'd like to be able to let others know I'm angry instead of swallowing my anger, as I do now."
- "I want to be able to challenge my brother when he takes the car whenever he wants to on weekends just because he's a year older than I am."

You're much more likely to change your behavior if what you're trying to change can be observed by you and by others and if instances of new behavior can even be counted.

- "I confronted my brother twice last week on using my clothes without asking me first. This is two times more than I would usually confront him. I wasn't aggressive. And I like being more assertive."
- "This week I called three people up at different times and asked them to do something socially, going to movies and things like that. Usually I sit and wait to be called, and, when the calls don't come, I feel left out."

320

Concreteness cannot be overstressed. If you say "I'm shy and want to do something about it," you won't do anything until you get a concrete picture of the ways in which you're shy and of the interpersonal behaviors you'd like to substitute for them.

Accentuating the Positive

Your goal isn't merely to get rid of some behaviors you don't like; that leads to a vacuum. Even if one of your principal goals is to get rid of some undesirable behavior, the change process should include putting something positive in its place. Never just decide *not* to do something.

- "One way in which I'm passive is to come on as a kind of 'yes man.' Whenever anyone I care about says anything, I agree with him or her. I say 'Yeah' or nod my head. I see now that I'm just trying to get on people's good side. I don't want to be rejected. I want to stop being a 'yes man.' "

This is good as far as it goes, but it doesn't say what kinds of behaviors you'd like to substitute for your 'yes-man' approach.

- "One way to get out of being a 'yes man' is to start by using basic understanding. I won't start by agreeing with what other people say. I'll start by understanding what they say. That way, I'll still show respect for others without just trying to get on their good side."

If you keep your attention on what you're doing wrong and merely try to eliminate negative behaviors, you'll establish a poor climate for changing your behavior. It's too easy to get down on yourself. If you accentuate the positive, however, you'll be confronting your strengths rather than just your weaknesses.

Rewarding Yourself

One of the most important change-of-behavior principles is that behaviors that are rewarded tend to be repeated and become part of our behavior patterns. Therefore, if you're trying to get rid of a behavior, reward yourself in some way when you stop engaging in that behavior. For instance, if you want to be assertive rather than aggressive, reward yourself when you use humor that isn't biting or sarcastic. Perhaps the best reward is the more positive response that you'll receive from others. They'll enjoy both you and your humor more, and they'll like to be with you more. However, you can also reward yourself in other ways. For example, you can say to yourself "If I control my sarcasm this afternoon, I'll let myself read a couple of chapters in a novel I want to read this evening." However, the more closely the reward is connected with the actual behavior you want to change, the better it is.

If you're trying to acquire a certain behavior, the process is the same: reward yourself when you engage in that behavior. For instance, you might say to yourself that you want to be more assertive in a certain way: "When my wife and children avoid touchy subjects at home, such as responsibility for keeping the house and yard and garage clean, I want to ask them directly to talk about who should do what and when." Again, the best reward comes from the more direct and open communication with your wife and children and from the fact that the work around the house gets done without a lot of fighting. However, you can also give yourself other rewards: "Everytime I challenge my wife or one of the kids to talk directly about taking responsibility for cleaning up—and do so without hostility—I'll reward myself by letting myself work for a half hour in the hobby shop." Good interpersonal interaction should ultimately be rewarding in itself, but you may find it helpful to give yourself other kinds of rewards until your new interpersonal behaviors are firmly established.

Since behavior can be closely connected to the rewards that it produces, you can ask yourself at the beginning of any change-of-behavior program "What is the payoff if I change my behavior?" This isn't as crass as it might sound. Even being kind to others has a payoff or reward in it. You find that you feel better when you go out of your way to do things for others, or it's rewarding to know that helping others fits in with your human or your religious values.

Biting Off Only What You Can Chew

Once you determine the behaviors you want to add to or subtract from your interpersonal style (I've been concentrating on assertiveness in the preceding examples), make sure that you're not trying to accomplish too much. Don't try to do everything at once.

Try to establish goals that are limited, possible, concrete, and workable. For instance, if you're working at becoming more assertive in your training group, your immediate goals might be stated somewhat as follows.

> "By being assertive I mean the following:
> - "I want to attend more carefully to what others are saying. I let myself sit back and even slouch. My body is a good indication of what's happening to me inside. I'll watch my posture and my physical involvement with the other members of the group. I'll ask them to remind me if they see me slipping away physically.
> - "I want to increase the number of times I respond with basic understanding. I don't do this spontaneously. I don't even think of doing it. For a while, this will be my basic response to others.
> - "My relationship with Gene isn't poor, but there's very little happening between him and me. I want to use relationship immediacy with him. I want to find out what causes us to be so bland with each other."

Here you specify the behaviors you want to increase, such as basic understanding, and the behaviors you want to decrease, such as slouching. Your goals are meaningful, but they aren't an attempt to become assertive in one leap.

Point A: Knowing Where You're Starting From

Your goals are point B; they indicate where you want to go. Point A is your realization of where you are now. It's impossible for you to determine whether you've changed or not unless you know what you're like to begin with. Therefore, watch yourself for a while, without trying to change. For instance, with respect to the three goals listed in the preceding example, you discover:

- that you end up in a slouching position three or four times per meeting,
- that you never respond with basic understanding unless someone asks you to, and
- that you initiate conversations with Gene only once per meeting.

323

Now you know where you are, and you'll be able to evaluate later on whether you're making progress. Progress in this case could mean:

- reducing slouching behavior to zero,
- responding with basic understanding to people in the group at least three or four times per meeting, and
- contacting Gene two or three times per meeting at a minimum.

Now you know both where you are (point A) and where you want to go (point B). Next you have to find out what keeps you from getting there and what helps you to get there.

Things That Help You and Things That Get in Your Way

As you proceed toward your goal, there are usually things that help you get there and things that stand in your way. It's good to know what these are, for then you can try to get rid of what stands in your way and to take advantage of what helps you. Thus, if you want to be more assertive in the group, you can make two lists.

A. *What helps me to be more assertive*
   - When I'm active in the group, I come away feeling good about myself.
   - People in the group encourage one another to respond with understanding. The person who does this is respected in the group.
   - The active people seem to be the most satisfied with the group experience. Those who work at assertiveness get the most out of the group.
   - When any group member brings up a problem he or she has relating to someone else in the group, the other members are very cooperative. They try to help the two people work out the relationship and generally offer constructive suggestions in a spirit of encouragement and support.

B. *What gets in the way of my being more assertive*
   - When I slouch, no one expects much of me, but when I attend, I'm expected to respond and generally be more

active. I go into the group knowing that I'll be under pressure.

- I'm still not very skilled in giving basic understanding. I know I need the practice, but I keep hesitating because I'm not good enough.
- Gene is a fairly passive person. I might not get much cooperation when I try to deal with "you-me" issues with him.

By writing down as concretely as possible what helps you and what gets in your way, you put all the cards on the table. Then you can begin to use what helps you and try to neutralize what gets in your way.

Finding Out What Emotions Are Getting in Your Way

If you're doing something you don't want to do (for instance, avoiding people who say they like you) or if you're not doing something you would like to do (not doing anything to establish a friendship with someone you like), it may be that certain emotions are getting in your way. Emotions are strong factors in keeping us from making the changes that we think we want in our lives.

Emotions and doing what you don't want to do. I've used the example of somebody who is a "yes man" and who doesn't want to be that way. If you're forever agreeing with others and doing nothing to make waves, there must be some kind of emotional payoff there for you. It may be that, if you always agree, people accept you, and you have feelings of security. Perhaps, when you do disagree with others, you find yourself feeling guilty for pushing your own opinions. If you keep agreeing with others, you don't experience guilt. Feelings of guilt and security, then, play a big part in your not changing. Consequently, if you want to change, you need to find other sources of security in interpersonal relating and to come to grips with your feelings of guilt. Assertiveness means, in part, not feeling guilty when you're getting your own needs met in ways that don't step on the rights of others.

Emotions and not doing what you would like to do. If you want to be more assertive in the ways we've discussed, you may find your emotions getting in the way of your changing. For instance, perhaps you want to provide more basic understanding for others, but fear and self-doubt keep you from doing it. You're afraid to speak up because you're not used to it, and you doubt your ability to communicate understanding well. Your fears and your lack of self-confidence keep you from doing what you want to do. Therefore, if you want to change, you have to come to grips with these emotions. One way of overcoming fears is first to realize intellectually that most of the horrible things we fear will result from our actions ("I'll try accurate understanding, and I'll fall on my face, and others will laugh at me and see me as stupid, and everything will be just horrible") usually don't happen. So often our fears are exaggerated. Therefore, if you can bring yourself to take a reasonable risk, you'll be rewarded with the discovery that all those horrible things that were supposed to happen didn't happen.

Lack of self-confidence is also very often based on self-defeating thinking. For example:

> "I can't try communicating understanding, because I'm not an expert in it, and that's what others expect of me—being perfect in the skill right from the start. Therefore, since I can't be perfect immediately, I won't even try."

You might not be this explicit with yourself, but this is often the kind of self-defeating thinking that goes on inside.

In summary, if you want to change, you should face the emotions that stand in the way of change, and you should face the kind of self-defeating thinking that keeps you locked into these emotions.

The avoiding process. All of us have a tendency to run away from whatever we see as harmful or unpleasant. This tendency, perhaps more than anything else, keeps us from using the resources we have. Let's take a look at why this is so. If you and I have a fight and leave each other still feeling very angry, the next time we see each other it's natural for us not to want to talk to each other. We may be at a large party where we talk to many different people, but both of us make sure that we avoid each other. The problem is that both of us find this avoiding *rewarding*. It's rewarding because we don't experience the unpleasantness that we experienced when we got into our argument. This is the key: *avoiding things can be rewarding*, and we tend to repeat behaviors that are rewarded. Thus, we tend to repeat the ways in which we avoid people and things. There comes a time when we no longer think directly about or even feel the reward. We're just in the habit of avoiding. We don't want to stir things up. We don't want to make things unpleasant again. The problem with avoiding is that we never give ourselves a chance to face and deal with our problems. If, instead of avoiding each other after our fight, you and I had given ourselves the opportunity of meeting again and, under more favorable emotional conditions, trying to understand each other and work out our differences, we might well have found that our new relationship was even *more* rewarding than avoiding each other had been. Again, it's a question of taking a risk. One of the values of the group is precisely that it can provide the kind of encouragement and support needed to take reasonable risks.

Moving Step by Step

One of the main reasons people give up change projects is that they try to do too much all at once. If you try to redo your interpersonal style overnight, you're bound to fail. Most of the things we don't like about ourselves are things that have been with us for a long time. If I come to dislike my passivity with others, the chances are that I have been a relatively passive person most of my life. I've *learned* my passivity, and now it's time to unlearn it. If I've spent years learning and practicing passivity, it isn't likely that I'll unlearn it and become assertive overnight.

The term psychologists use for the step-by-step approach to changing behavior is *shaping*. When you take gradual steps to change your behavior, you're shaping your behavior. The key in a change-of-behavior project is to start small, reward yourself for small successes, and keep linking smaller behaviors together until you achieve the larger behaviors you're interested in. If you want to be more assertive in the skill of immediacy, first of all get comfortable with the communication of basic understanding and with the other skills that go into good immediacy (self-disclosure, inviting the other person to examine an issue with you, and so on). In choosing issues to talk about, don't start with the most serious ones. Start with something that you can handle relatively easily. Start with something positive.

> *You:* "Tim, you and I seem to be similar in a few ways. For instance, you're very eager to learn these skills. And, if I'm not mistaken, you can see my enthusiasm also. That makes me want to work with you. It makes the work of the group easier for me, because I feel you're on my side."

327

Then, through immediacy transactions of greater depth and seriousness, you can gradually deepen your relationship with Tim. If you were to start in the following way, we could almost predict the results:

> *You:* "Tim, I like you very much. I'd really like to get very close to you. I know that you like me somewhat, but I'd like you to feel the same way about me that I feel about you."

This approach might scare anybody off. It's too much too soon. If you go too fast in a change program, you might scare yourself and scare others, become depressed by your failures, and give up. To get through a change-of-behavior program successfully, you need some feelings of success step by step.

Models

The group is a place where you can find models of some of the interpersonal behaviors you'd like to acquire. This, too, is one of the principles of change. If you can get a picture of the kind of behavior you want by seeing another person do it well, this picture might well be worth a thousand words of description in a book. For example, if someone in your group is good at being assertive in terms of the skill of immediacy ("you-me" talk), study his or her style. Adopt from it what you think would fit your style. You don't have to give up your own individuality in doing so. Give yourself permission to borrow parts of other people's styles, and then fit them together into your own unique style.

A Systematic Change Program

The last principle of behavioral change is to avoid the mistake of thinking that change will just happen. It won't. Change takes place most effectively if you go about it systematically—that is, if you put into practice the rules and principles I've been describing. We'll now look at an exercise in which you're asked to do a change-of-behavior program and to do it systematically. Remember what was said about the difference between living spontaneously and living haphazardly. Living spontaneously requires discipline. Being systematic in a change-of-behavior program also requires discipline.

328

Exercise 54
A Systematic Program for Changing Your
Interpersonal Behavior

What follows is, in simplified form, a program of self-change. Since you're working in the area of interpersonal relationships, and since you're trying to become more assertive in these relationships, choose some area of interpersonal relating that you would like to improve both inside and outside the group. For instance, you might want to increase the quantity and quality of your immediacy ("you-me") interchanges both in the group and outside the group.

Here are the steps for your change program. Keep a notebook on the entire change program, indicating what you do and with what success at each stage.

1. *Choose a general area* you'd like to work on (your interpersonal style) and how you'd like to work in that area (for example, increasing your level of assertiveness in some interpersonal area).
2. *Limit your program;* that is, choose some specific, concrete goal (for instance, to increase the quantity and quality of your immediacy interchanges both inside and outside the group). Use the feedback you've already received in the group to pick a topic. If you've been told that you seem quite reluctant to disclose yourself very deeply, this is an area you might experiment with. Since you have been examining your interpersonal style and have begun to see certain patterns you may not like, use this information, too, to pick a topic. For instance, you may notice that, when others come on strong in their interpersonal styles, you withdraw. You don't initiate conversations, you tend to agree with them even when you really don't, and so forth. Seeing this pattern in yourself will help you pick the kind of assertiveness you'd like to develop.
3. *Make your goal concrete.* Talk about the specific behaviors that you'd like to engage in. This will help you limit the area in which you're working. Don't bite off more than you can chew. In this first project, try something that will make a difference in your life but that is

329

quite possible for you to do. For instance, you might want to increase the number of times you respond to others inside or outside the group with basic, accurate understanding. Write clearly and distinctly where you want to go; call this point B.

4. *Get a good feeling for point A.* Remember that point A is your starting point, what you're like now. Review your behavior in the group and in interpersonal situations outside the group. Understand carefully what your behavior is like now. Perhaps you find that you never initiate immediacy interchanges, either in the group or in your everyday life. You know this skill; you've practiced it in the group when you've been asked to do so; but you don't do it spontaneously.

5. *Become very specific about point B.* If possible, make your goal numerical. For example, pick two people in the group and two people outside the group with whom you'd like to have immediacy interchanges. Decide how often you want to engage in "you-me" talk with them.

6. *List the things that will get in the way of your goal.* What—especially what emotions—will keep you from your goal? Are you lazy? Make a list of the things that will make it hard for you to strive for your goal. It may be that one of the people with whom you'd like to be more immediate is a very passive person. You're not sure whether you're going to get a response. You're not sure whether you want to put effort into contacting that person if there isn't going to be any payoff.

7. *List the things that will help you get to your goal.* What can you count on to help you get to your goal? Are the people you're interested in getting closer to also interested in getting closer to you? Are they developing useful interpersonal skills? Find out what you have going for you.

8. *Review your motivation.* Ask yourself how deeply you're committed to this change project. Is it something you really want? Are you willing to work for it? If you're only halfhearted in your desire to change, it might be better to put off the change project until you deal with your own lack of motivation.

9. *Spell out a step-by-step program.* You now know both point A and point B and can ask yourself how you can

move step by step from point A to point B. What is the first step you should take? Is it too big? Can you divide it into two smaller steps? For example, choose the person who is most likely to respond well to you as the first one with whom to have an immediacy exchange. Don't build failure into your program by trying to move too quickly, but don't move so slowly that you lose interest in the project.

10. *Decide what rewards you expect from the change program.* Not only should you know what reward to expect at the end (for instance, increased enjoyment with your friends if, when the occasion calls for it, you can be immediate with them), but you should also know what's going to keep you working. Rewards keep our motivation up. What rewards to you expect from your first step? Do you intend to have your raw enthusiasm or raw determination carry you through the whole project? Is that realistic? For instance, if the first thing you need to do is to get better at the communication of basic understanding, then just getting better at this skill might be a reward in itself. Have you arranged for outside rewards in case the behaviors you're engaging in aren't rewarding enough in themselves?

11. *Evaluate yourself.* One reason for concreteness in the beginning of your project (for instance, clearly spelling out both point A and point B) is to give you a clear picture and clear standards, so that you'll be able to find out whether you're making progress or not. You need both self-evaluation and feedback from others, but none of this helps unless your standards are clear. For instance, you can say "This week in the group I initiated two immediacy interchanges with John. We both found them rewarding. I feel much closer to John." Evaluation isn't something to be done just at the end of an entire project. Evaluate yourself at each step of the project. If you want to improve the quality of your communication of basic understanding as a step in a program to initiate more immediacy with others, find out from others whether you're getting better in basic understanding.

Transferring What You Learn in the Lab to Your Everyday Life

This book has been written from the point of view of your participation in a training laboratory. However, the substance of this book and of what you do in the lab applies to your life outside the lab, too. In other words, the lab isn't a game; there should be dialogue between what you do in the lab and your relationships outside the lab. Let's consider some of the factors involved in transferring your learning from the lab to your day-to-day life.

• *Motivation.* Most of what you learn in the lab will transfer from the lab to your daily life *if you want it to*. More concretely, this means that the transfer will take place if what you learn in the lab *works* outside and if you feel *rewarded* for using your new skills. The lab suggests that you live more intensively with others through both support and challenge. If this scares you, then what you've learned might stay locked up inside you.

• *What you do in the lab.* The more you learn in the lab, the more you can transfer to your day-to-day life outside. While half-learned skills may make your relationships with others worse—for instance, if you learn to confront but without understanding—fully learned skills are rewarding not only for you but for those with whom you live, work, and play.

• *Support for your learning outside the group.* If you and your fellow group members have worked hard, then you have probably developed a good climate of support in your group. It may be that at first you won't find the same kind of support outside. It's easy to run into people who don't like closeness and who don't want to live very intensively. On the other hand, there are many people who are looking for precisely these things in their lives. Outside the lab, then, neither apologize for your skills nor beat others over the head with them. Look for those who are interested in relating more deeply. Then you'll find the support you need to keep challenging yourself.

• *Moving slowly.* The alternative to inflicting yourself, your newly won skills, and your improved interpersonal style on others is to let the changes in your style speak slowly for themselves. If you begin by using more understanding in a natural, relaxed way, this might well be an important difference in your style—a gentle, useful, and usually appreciated difference. People are usually capable of accepting and appreciating small rather than large changes.

A Concluding Note

There's no magic in the pages of this book. There is only the suggestion of hard work. It's impossible to write a book on almost any subject without giving the impression that this is the most important subject in the world. Interpersonal skills—one-to-one and in groups—are extremely important in life, but there are other important things: work, solving problems, poetry, the poor, art, the sick, music, the changes we need in our institutions and in our social order, religion, politics, vacations, learning, romance, fun, books, friends, pain, career, success, failure, the rest of the world, and dying. What you learn here in this lab, however, can help you to invest yourself in many of the other parts of your life more creatively.

Bibliography

Alberti, R. E., & Emmons, M. L. *Your perfect right: A guide to assertive behavior* (2nd ed.). San Luis Obispo, Calif.: Impact, 1974.

Alberti, R. E., & Emmons, M. L. *Stand up, speak out, talk back.* New York: Pocket Books, 1975.

Bennett, C. C. What price privacy? *American Psychologist,* 1967, *22,* 371–376.

Berenson, B. G., & Mitchell, K. M. *Confrontation: For better or worse.* Amherst, Mass.: Human Resource Development Press, 1974.

Brandon, N. *The psychology of self-esteem.* New York: Bantam Books, 1969.

Bullmer, K. *The art of empathy: A manual for improving accuracy of interpersonal perception.* New York: Human Sciences Press, 1975.

Carkhuff, R. R. *The art of problem solving.* Amherst, Mass.: Human Resource Development Press, 1973.

Carkhuff, R. R. *How to help yourself: The art of program development.* Amherst, Mass.: Human Resource Development Press, 1974.

Culbert, S. A. *The interpersonal process of self-disclosure: It takes two to see one.* Fairfax, Va.: Learning Resources Corp./NTL, 1967.

Egan, G. *Encounter: Group processes for interpersonal growth.* Monterey, Calif.: Brooks/Cole, 1970.

Egan, G. (Ed.). *Encounter groups: Basic readings.* Monterey, Calif.: Brooks/Cole, 1971.

Egan, G. *Face to face: The small-group experience and interpersonal growth.* Monterey, Calif.: Brooks/Cole, 1973.

Egan, G. *Exercises in helping skills: A training manual to accompany* The Skilled Helper. Monterey, Calif.: Brooks/Cole, 1975.

Egan, G. *The skilled helper.* Monterey, Calif.: Brooks/Cole, 1975.

Egan, G. *Interpersonal living.* Monterey, Calif.: Brooks/Cole, 1976.

Gazda, G. M. *Human relations development.* Boston: Allyn & Bacon, 1973.

Harris, T. *I'm OK-you're OK: A practical guide to transactional analysis.* New York: Harper & Row, 1969.

335

Howard, J. *Please touch: A guided tour of the human potential move-ment.* New York: McGraw-Hill, 1970.

James, M., & Jongeward, D. *Born to win: Transactional analysis with Gestalt experiments.* Reading, Mass.: Addison-Wesley, 1971.

Johnson, D. W. *Reaching out: Interpersonal effectiveness and self-actualization.* Englewood Cliffs, N. J.: Prentice-Hall, 1972.

Jourard, S. M. *The transparent self* (Rev. ed.). New York: Van Nos-trand Reinhold, 1971.

Keen, S., & Fox, A. V. *Telling your story: A guide to who you are and who you can be.* New York: Doubleday, 1973. (Also New American Library, 1974.)

King, S. W. *Communication and social influence.* Reading, Mass.: Addison-Wesley, 1975.

Lazarus, A., & Fay, A. *I can if I want to.* New York: William Morrow, 1975.

Liberman, R. P., King, L. W., DeRisi, W. J., & McCann, M. *Personal effectiveness: Guiding people to assert themselves and improve their social skills.* Champaign, Ill.: Research Press, 1975.

Lynd, H. M. *On shame and the search for identity.* New York: Science Editions, 1958.

Maslow, A. *Toward a psychology of being* (2nd ed.). New York: Van Nostrand Reinhold, 1968.

Mayeroff, M. *On caring.* New York: Perennial Library (Harper & Row), 1971.

Mehrabian, A. *Silent messages.* Belmont, Calif.: Wadsworth, 1971.

Powell, J. *Why am I afraid to tell you who I am?* Niles, Ill.: Argus Communications, 1969.

Rogers, C. R. *On encounter groups.* New York: Harper & Row, 1970.

Simon, S. B. *Meeting yourself halfway: Thirty-one value clarification strategies for daily living.* Niles, Ill.: Argus Communications, 1974.

Simpson, C. K., & Hastings, W. J. *The castle of you: A personal growth workbook.* Dubuque, Iowa: Kendall/Hunt, 1974.

Steiner, C. *Scripts people live.* New York: Grove Press, 1974.

Wallen, J. L. Developing effective interpersonal communication. In R. W. Pace, B. D. Peterson, & T. R. Radcliffe (Eds.), *Communicating interpersonally.* Columbus, Ohio: Charles E. Merrill, 1973. Pp. 218–233.

Watson, D. L., & Tharp, R. G. *Self-directed behavior: Self-modification for personal adjustment.* Monterey, Calif.: Brooks/Cole, 1973.

Williams, R. L., & Long, J. D. *Toward a self-managed life style.* Boston, Mass.: Houghton Mifflin, 1975.

Appendix:
Suggested Responses
to Exercises 25-29 and 31

Exercise 25

1. hurt, put out, put down, angry, rejected
2. pleasantly surprised, happy, full of enthusiasm, energetic, satisfied
3. guilty, ashamed, sorry, eager to set things right
4. disturbed, anxious, annoyed, confused, fearful, on edge
5. relieved, surprised, good, off the hook
6. hesitant, confused, hurt, perplexed, sad
7. in a bind, torn two ways, uncomfortable, uneasy, like he doesn't know what to do
8. a little embarrassed, loving, caring, trusting, appreciative, attracted
9. depressed, self-doubting, down on herself, inferior, like running away
10. comfortable, at peace, relieved, more together

Exercise 26

1. . . . he gets such negative feedback from someone he wants to get to know.
2. . . . he's actually learning skills that he can use in real life.
3. . . . he's been coming on too strong.
4. . . . he sees silence as covering over serious issues that should be faced.
5. . . . she finds out that she's not really a "bad guy" in Carl's eyes at all.
6. . . . there is a lack of solid give-and-take in his relationship with Don.

7. . . . he wants a deeper relationship with Laura, but he's not sure that she feels the same way.
8. . . . he likes being in an equal and deepening relationship with Carl.
9. . . . she sees herself as so inferior to the other members of the group.
10. . . . her fight with Maureen didn't destroy their relationship but merely cleared the air.

Exercise 27

1. angry, very annoyed, boiling mad, really ticked off, exasperated, like blowing up, very upset, like blowing your lid, almost like ditching this whole group experience, furious
2. good, relieved, like you're finally getting somewhere, satisfied, content, at peace, relaxed, settled, like you're in a good place
3. pleasantly surprised, grateful, refreshed, challenged but supported, like something good has happened to you, really pleased
4. tired, like you've had it, annoyed, at your wit's end, fed up, like taking a break, punchy, exhausted, put out, like nothing's happening, like we're getting nowhere, tuned out
5. ambivalent, cautious, hesitant, unsure of yourself, like you both want to and don't want to, like you're not sure what you want, in a fog, up in the air, confused
6. hesitant about getting in deeper, suspicious, up in the air, confused, at sea, cautious, reluctant to involve yourself, uneasy, edgy
7. annoyed, on the ropes, up against the wall, picked on, pushed too hard, like blowing off some steam, mistreated, frustrated
8. like you've said more than you want to, anxious, exposed, like shutting down, defenseless, uneasy, embarrassed, pushed, irritated, like you're already in too deep, ashamed, like taking a step back
9. determined, like you've got a good head of steam up, like moving ahead, resolved to get what you've paid for, insistent, like pushing ahead
10. exhausted, drained, like you've put in a very heavy day, tired but still willing to work, determined to work in spite of a lousy day, like you've had enough for today, at a low ebb, like too much has happened, played out, on the ropes

Exercise 28

1. . . . everybody here is so wrapped up in his own needs.
2. . . . you've finally made contact with each other.
3. . . . you never expected to hear such a clear and straightforward message from me.
4. . . . pushing you right now is not really helping anything.
5. . . . you're still not too sure how much you want to invest yourself here.
6. . . . you want to know just where you stand with me before getting involved with me.
7. . . . you see me just picking away at you without really being "with" you.
8. . . . you think you've told too much too soon.
9. . . . you see a lot of opportunities here, and you're determined to take advantage of them.
10. . . . you've had a heavy day, even though it's over now.

Exercise 29

1. a. "You feel angry because you think it's useless to compare yourself to others, and now I'm doing it for you."
   b. "I just make things worse when I compare you to others. This really annoys you, because you do enough of that yourself."
2. a. "You feel down on yourself because you keep comparing yourself to me, and this makes you feel only more inadequate."
   b. "It's a real downer for you when you keep seeing me as competent and yourself as incompetent."
3. a. "You feel pretty good about yourself because the skills you've learned have helped your self-confidence a lot."
   b. "Learning these interpersonal skills has helped your self-esteem a lot. And that's a good feeling!"
4. a. "You feel uneasy about talking to me because you think of me as someone really beyond the ordinary."
   b. "Once you put me on a pedestal, well, maybe it's too scary then to talk to someone out of the ordinary."
5. a. "You feel depressed because it's important to you that I call you, and I just don't."

    b. "Waiting for the phone to ring can really get depressing. And I just haven't called."

6. a. "You feel pressured into sharing yourself with me because I've shared something rather intimate about myself with you."

    b. "Sharing how I feel about myself sexually can put you on the spot to do the same. But you don't want to be pressured."

7. a. "You feel confused because you're not sure how I really stand on the issue of going to the lake."

    b. "I guess my moodiness makes you wonder whether I'm put out about not going to the lake or whether it's something else."

8. a. "You feel guilty because you fear that you've violated some of the norms of the group or because you fear that we think you have."

    b. "You think that you may have violated the trust of the group by going out with Steve. And that seems to be upsetting you right now."

9. a. "You feel bored and frustrated because the level of risk-taking here, including your own, is just too low."

    b. "The group experience has been pretty 'blah' for you. And you're willing to take your share of the responsibility."

10. a. "You feel pretty down because your troubles in school have begun to mess up your relationships with your friends."

    b. "Feeling messed up in school is bad enough. But when it also messes things up with your friends—that's really bad."

Exercise 31

1. a. minus: interpretations, playing psychologist
    b. minus: judgmental
    c. minus: parental, premature and poor confrontation, advice giving
    d. plus
2. a. minus: inaccurate, judgmental
    b. minus: a non-response, no understanding at all
    c. minus: inappropriate question, ignoring speaker's feelings
    d. minus: ignores feelings, poorly worded and premature confrontation, doesn't get into "world" of speaker

3. a. minus: cliché, indirectly confrontational and judgmental
   b. minus: interpretation, inappropriate confrontation, patronizing
   c. plus
   d. minus: no accurate understanding, self-disclosure without first understanding the other, somewhat patronizing and too "nice"
4. a. minus: cliché
   b. minus: patronizing, ignores feelings of speaker
   c. minus: inaccurate understanding
   d. minus: defensive, accusatory
5. a. minus: changes subject, ignores what speaker has said, especially speaker's feelings; self-centered, accusatory
   b. minus: inappropriate question, sidetracking
   c. plus
   d. minus: parroting
6. a. minus: approval rather than understanding, takes sides
   b. minus: confrontation without understanding, accusation, judgmental
   c. minus: long-winded
   d. minus: inadequate response, ignoring what speaker has said
7. a. minus: inadequate response, cliché, ignores speaker's feelings
   b. minus: doesn't deal with central issue, ignores real feelings and content
   c. plus
   d. minus: ignores feelings and content, pushes own needs

Index

344